Biotech's

DICTIONARY Of Global Positioning System

Compiled & Edited by
Pradeep Kumar

BIOTECH

2007

BIOTECH BOOKS

DELHI - 110035

Biotech's

DICTIONARY *Of* Global Positioning System

Compiled & Edited by
Pradeep Kumar

First Indian Edition 2007

ISBN 81-7622-180-5

Published by : BIOTECH BOOKS
1123/74, Tri Nagar,
Delhi - 110 035
Phone: 27382765
e-mail: biotechbooks@yahoo.co.in

Showroom : 4762-63/23, Ansari Road, Darya Ganj,
New Delhi - 110 002
Phone: 23245578, 23244987

Printed at : Chawla Offset Printers
New Delhi - 110 052

PRINTED IN INDIA

Preface

Global Positioning System, usually called GPS, is the only fully-functional satellite navigation system. A constellation of more than two dozen GPS satellites broadcasts precise timing signals by radio to GPS receivers, allowing them to accurately determine their location (longitude, latitude, and altitude) in any weather, day or night, anywhere on Earth.

GPS has become a vital global utility, indispensable for modern navigation on land, sea, and air around the world, as well as an important tool for map-making, and land surveying. GPS also provides an extremely precise time reference, required for telecommunications and some scientific research, including the study of earthquakes.

The first in a series of next-generation GPS satellites was added to the constellation, offering several new capabilities, including a second civilian GPS signal called L2C for enhanced accuracy and reliability. Additional next-generation satellites will increase coverage of L2C and add a third and fourth civilian signal to the system, as well as advanced military capabilities. The Wide-Area Augmentation System, available since August 2000, increases the accuracy of GPS signals to within 2 meters (6 ft) for compatible receivers. GPS accuracy can be improved further, to about 1 cm (half an inch) over short distances, using techniques such as Differential GPS.

GPS allows accurate targeting of various military weapons including, cruise missiles and precision-guided munitions, as well as improved command and control of forces through improved locational awareness. The satellites also carry nuclear detonation detectors, which form a major portion of the United States Nuclear Detonation Detection System. Civilian GPS receivers are required to have limits on the velocities and altitudes at which they will report coordinates; this is to prevent them from being used to create improvised missiles.

GPS is used by people around the world as a navigation aid in cars, airplanes, and ships. The system can also be used by computer controlled harvesters, mine trucks and other vehicles. Hand-held GPS receivers can be used by mountain climbers and hikers. Glider pilots use the logged signal to verify their arrival at turn points in competitions. Low cost GPS receivers are often combined in a bundle with a PDA, car computer, or vehicle tracking system. GPS equipment is even available for the visually impaired.

The book covers all the latest terms in technology, applications, and systems. The definition includes explore the integration of GPS with vehicles and cellular telephones, new classes of satellite broadcast signals, the emerging GALILEO system, and new developments in the GPS marketplace. This single-source dictionary provides a quick overview of GPS essentials, an in-depth examination of advanced technical topics, and a review of emerging trends in the GPS industry. Engineers can use this book to use GPS terms and integrate them into navigational and communications equipment. Executives can turn to this book to determine how technology is affecting markets and how best to invest their companies' resources. This dictionary also serves as a handy resource for engineering students looking to advance their studies and careers in GPS.

- **10base-2**
  Computers, Networking: An Ethernet LAN 10 Mbps baseband transmission standard using 50-ohm coaxial cable (RG-58 A/U) with BNC connectors connected to network cards via T-connectors. Unused connections must be terminated using 50-ohm terminators. The maximum cable length is 185 meters. Also called Thinnet.

- **10base-36**
  Computers, Networking: An Ethernet LAN 10 Mbps baseband transmission standard using broadband 50-ohm coaxial cable (RG-58 A/U) with BNC connectors connected to network cards via T-connectors. Unused connections must be terminated using 50-ohm terminators. The maximum cable length is 3600 meters.

- **10base-5**
  Computers, Networking: An Ethernet LAN 10 Mbps baseband transmission standard using 50-ohm coaxial cable (RG-58 A/U) with BNC connectors connected to network cards via T-connectors. Unused connections must be terminated using 50-ohm terminators. The maximum cable length is 500 meters. Also called Thicknnet.

- **10base-f**
  Computers, Networking: An Ethernet LAN 10 Mbps baseband transmission standard using fiber-optic cable.

- **10base-t**
  Computers, Networking: An Ethernet LAN 10 Mbps baseband transmission standard using twisted-pair telephone wire.

- **16:9**
  Video: The aspect ratio for widescreen DTV video.

- **2 d seismic**
  acoustic measuring technology which uses single ship-towed hydrophone streamers. This technique generates a 2D cross-section of the deep seabed and is used primarily when initially reconnoitring for the presence of oil or gas reservoirs.

- **2.5g**
  Acronym: 2.5 Generation Wireless. Telephony: A term used to describe wireless telecommunications technology that uses circuit-switched networks to approach the capabilities offered by 3G networks. 2.5G is an interim stage in wireless telecommunications development.

- **2d mode**
  GPS Acronym: 2 Dimensional Mode. GPS: A two-dimensional GPS position fix that includes only horizontal coordinates (no elevation). It requires a minimum of three visible satellites.

- **2d operating mode**
A two-dimensional GPS position fix that includes only horizontal coordinates (no GPS elevation). It requires a minimum of three visible satellites.

- **2drms**
A position accuracy measure defined as twice the RMS of the horizontal error. This approximately corresponds to the 95% confidence interval, or "two sigma" value. For example, under the policy of "Selective Availability" GPS Absolute Positioning accuracy is claimed to be 100m 2drms, which means that approximately 95% of the horizontal position solutions will be within 100m of the correct value; Twice-distance Root Mean Square, a horizontal measure of accuracy representing the radius of a circle within which the true value lies at least 95 per cent of the time.

- **2g**
Acronym: 2nd Generation Wireless. Telephony: A term used to describe digital cellular telephones that use circuit-switched networks.

- **3 d seismic**
accoustic measuring technology which uses multiple ship-towed long hydrophone streamers. This technique generates a 3D model of the deep seabed and is used to locate and analyse oil and gas reservoirs.

- **3 diq (3d integrated quantitative)**
technology for the development of integrated (geology, geophysics, reservoir engineering) quantitative oil and gas reservoir models; these models are used to optimise the risks, costs and efficiency of oil and gas field development and production.

- **300-ohm**
Cables: The standard impedance for twin-lead antenna connections found on older antennas and televisions.

- **3d mode**
Acronym: 3 Dimensional Mode. GPS: A three-dimensional GPS position fix that includes horizontal coordinates plus elevation. It requires a minimum of four visible satellites; 3 Dimensional Mode. A three-dimensional GPS position fix that includes horizontal coordinates plus elevation. It requires a minimum of four visible satellites.

- **3d operating mode**
A three-dimensional GPS position fix that includes horizontal coordinates, plus elevation. It requires a minimum of four visible satellites.

- **3g**
Acronym: 3rd Generation Wireless. Telephony: A term used to describe digital cellular

telephones that use packet-switched networks.

- **4:3**
  Video: The aspect ratio for NTSC video.

- **4g**
  Acronym: 4th Generation Wireless. Telephony: A term used to describe the next generation of wireless telecommunications, expected to deliver wireless multimedia content anywhere, anytime and to any device.

- **5.1 sound**
  Audio: A sound encoding method consisting of five full-bandwidth channels (front right, front left, center, rear right and rear left) plus ".1", a limited-bandwidth subwoofer channel.

- **50-ohm**
  Cables: The standard impedance for coax computer networking cable.

- **75-ohm**
  Cables: The standard impedance for coax audio/video cable used between antennas and receivers/ television.

- **8 mm**
  Video: A videotape format used in older and some current camcorders.

- **absolute positioning**
  A mode in which a position is identified with respect to a well-defined coordinate system, commonly a geocentric system (i.e., a system whose point of origin coincides with the center of mass of the earth); Mode in which a position is determined, using a single receiver, with respect to a well-defined coordinate system, typically a Geocentric system (i.e., a system whose point of origin coincides with the centre of mass of the earth). Also referred to as Point Positioning, or Single Receiver Positioning.

- **acc (adaptive cruise control)**
  Cruise control system that uses sensors to automatically maintain a safe distance from the vehicle ahead, adapting to speed changes, etc.

- **access charge**
  A fee charged subscribers or other telephone companies by a local exchange carrier for the use of its local exchange networks.

- **accuracy**
  A measure of how close an estimate of a GPS position is to the true location; The degree of closeness of a measurement result to its true value. The degree of conformance between the estimated or measured position, time, and/or velocity of a GPS receiver and its true time, position, and/or velocity as compared with a constant standard. Radio navigation system accuracy is usually presented as a statistical measure of system error.

- **accuracy relative**
  The accuracy with which a user can measure position relative to that of another user of the same navigation system at the same time

- **accuracy statistic**
  The industry standard definition of stataic accuracy is the standard deviation of error if a point set by a receiver was tested in a fix position for 24 hours.

- **acquisition**
  The ability to find and lock on to satellite signals for ranging.

- **acquisition of signal(aos)**
  The time at which a radio signal is first heard from a satellite, usually just after it rises above the horizon.

- **acquisition time**
  The time it takes a GPS receiver to acquire satellite signals and determine the initial position.

- **active antenna**
  An antenna that amplifies the GPS signal before sending it to the receiver.

- **active leg**
  A "segment" is that portion of a route between any two waypoints in the route.

- **active log**
  The segment of a route currently being traveled.

- **acts (advanced communications technology satellite)**
  The NASA experimental satellite project which demonstrated the use of the Ka Band (30/20 GHz) services; on-board processing with multiple beams, electronic hopping antenna beams, and dynamic adjuctment of power to cope with rain attenuation. The experimental network supported low to medium rate mobile services to high data rate HDTV fixed services.

- **actual sonar returns**
  With Advanced Fish I.D.™ off, Lowrance fish finders display actual sonar returns, which show likely game fish targets as fish arches. Actual returns also reveal important underwater detail like thermoclines, baitfish and fish movement.

- **adas (advanced driver assistance systems)**
  In-vehicle technologies designed to improve vehicle safety by aiding the driver, such as collision avoidance, curve warning, distance control, etc.

- **additional secondary factor (asf)**
  Error propagation time caused by radiowaves slowing down as they pass over different terrain. This is in addition to slowing down as radiowaves pass over sea water.

- **adrg**
  ARC-digitized raster graphics. A United States military map format for providing maps on CD.

- **advanced fish symbol i.d.™**
  Also called Fish I.D., it's what many anglers say is the easiest way to identify fish targets on their Lowrance units and is ideal for fishing shallow water. With Fish I.D. on, it automatically interprets sonar echos and displays them as fish symbols. The fish symbols are displayed in up to four different sizes to match the relative size of the target echo. With Fish I.D. off, and showing actual sonar returns, targets are displayed as fish arches.

- **advanced signal processing (asp™)**
  Available on all Lowrance fish finders, ASP™ helps you easily spend more time fishing, and less time adjusting sonar sensitivity. With ASP™ on, it automatically fine-tunes your sonar settings for the clearest, most accurate picture of the underwater world. As water conditions change, ASP™ filters out "clutter" that interferes with sonar signals...such as boat speed and suspended particles like algae and plankton...for the best display of fish targets under your boat.

- **advanced transportation management systems (atms)**
  The gathering of information through various means and compiling it to disseminate the information to the public through Advanced Traveler Information Systems.

- **advanced traveler information systems**
  The process of disseminating information collected by highway sensors to all drivers for pretrip route planning.

- **aerotriangulation (phototriangulation)**
  Aerotriangulation, also known as phototriangulation, is a complex process vital to aerial photogrammetry. It involves extending vertical and/or horizontal point control so that the measurements of angles and/or distances displayed on an overlaid photograph are made relative to a spatial solution using photographic perspective principles, correcting for photographic distortions. Aerotriangulation consists of mathematically extending the vectors/angles of a triangular pattern of known reference points on or near the designated photo block terrain upward through a rectangle representing the area of the photo-block as seen by the camera's optical center. In this way, the three-point terrain triangle and the three-point triangle of the camera's eye within the photographic frame are made analogous.

- **afe**

Atlas Florae Europaeae. The AFE grid is modified from the Military Grid Reference System (MGRS). The MGRS itself is an alphanumeric version of a numerical UTM (Universal Transverse Mercator) or UPS (Universal Polar Stereographic) grid coordinate; The AFE grid is modified from the Military Grid Reference System (MGRS).

- **afs (advanced frontline system)**

Concept of intelligent lighting, according to curves, weather and speed.

- **agonic line**

An imaginary line on the earth's surface connecting points where the magnetic declination is zero.

- **agps**

Assisted Global Positioning System. A land station assists GPS in acquiring position.

- **ahs (automated highway system)**

Research initiative aimed at developing automatic guidance on motorways for vehicles using cruise control.

- **aid (automatic incident detection)**

Radar- or video-controlled system enabling rapid location of an incident or accident on tunnels and motorways.

- **alarms**

Alarms are designed to alert you to certain sonar "events" by sounding tones. With alarms, you don't need to constantly look at the display, so you spend more time fishing. Lowrance features three main types of alarms, and all are selectable on/off by the user. The FISH ALARM alerts you when a fish symbol appears on the screen, with different tones for different sizes of fish targets. The ZONE ALARM lets you set an underwater depth zone (e.g., 21 ft. to 53 ft.) and sounds an alarm tone when targets are detected in that zone.

- **alert (advice and problem location for European road traffic)**

Protocol used for traffic messages by RDS-TMC.

- **algorithm**

A special method used to solve a certain type of mathematical problem; A procedure or formula for solving a mathematical problem in a finite number of steps. Extremely complex algorithms or sets of algorithms are used to simplify, modify or predict data in the digital manipulation of information. Often, algorithms which are developed for digital applications are proprietary and are an important aspect of defining significant differences among the various digital compression and codec (coding/decoding) standards.

- **almanac**
A data file that contains orbit information on all satellites, clock corrections, and atmospheric delay parameters. It is transmitted by a GPS satellite to a GPS receiver, where it facilitates rapid satellite vehicle acquisition within GPS receivers. Almanac data must be acquired before GPS navigation can begin; A data file that contains the approximate orbit information of all satellites, which is transmitted by each satellite within its Navigation Message. It is transmitted by a GPS satellite to a GPS receiver, where it facilitates rapid satellite signal acquisition within GPS receivers. Almanac data is kept current within a GPS receiver to facilitate "hot starts" by permitting the Doppler Shift of each satellite signal to be determined and configuring each tracking channel for this Doppler-shifted carrier frequency; The Almanac is a file which contains positional information for all of the GPS satellites. The Almanac is used by the GPS receiver to determine which satellites to track, and can also be used for mission planning.

- **almanac data**
Information transmitted by each satellite on the orbits and state (health) of every satellite in the GPS constellation. Almanac data allows the GPS receiver to rapidly acquire satellites shortly after it is turned on; Orbital position data for each GPS satellite. Considered not to be very accurate as the data are valid for over one month.

- **altimeter**
An instrument for determining elevation, especially an aneroid barometer used in aircraft that senses pressure changes accompanying changes in altitude. The Garmin® eTrex® Vista and Summit models contain a basic GPS with a built-in barometric altimeter; An instrument that measures altitude or elevation with respect to a reference level, usually mean sea level, by means of air pressure.

- **altitude**
The current elevation above sea level.

- **am (amplitude modulation)**
Modulation which is accomplished by varying the amplitude (height of the carrier wave) of the carrier signal in accordance with the information to be transmitted.

- **am (asset management)**
a management system that ensures the efficient use of business equipment such as ships, measuring equipment, etc.

- **ambiguity**
Carrier phase measurements can only be made in relation to a cycle

or wavelength of the L1 or L2 carrier waves because it is impossible to discriminate different carrier cycles (they are all "sine waves" if one ignores the modulated messages and PRN codes). Integrated carrier phase measurements may be made by those receivers intended for carrier phase-based positioning. In this case the change in receiver-satellite distance can be measured by counting the number of whole wavelengths since initial signal lock-on and adding the instantaneous fractional phase measurement. However, such a measurement is a biased range or distance measurement because the initial number of whole (integer) wavelengths in the receiver-satellite distance is unknown. This unknown value is referred to as the "ambiguity". It is different for the different satellites, and different for the L1 and L2 measurements. It is, however, a constant if signal tracking continues uninterrupted through an observation session. If there is signal blockage, then a "cycle slip" occurs, causing the new ambiguity after the cycle slip to be different from the value before. Cycle slip repair therefore restores the continuity of carrier cycle counts and ensures that there is only one ambiguity for each satellite-receiver pair; Integer bias term, Cycle ambiguity the unknown number of whole wavelengths of the carrier signal contained in an unbroken set of measurements from a single satellite at a single receiver; The initial bias in a carrier-phase observation of an arbitrary number of cycles. The initial phase measurement made when a GPS receiver first locks onto a GPS signal is ambiguous by an integer number of cycles because the receiver has no way of knowing the exact number of carrier wave cycles between the satellite and the receiver. This ambiguity, which remains constant as long as the receiver remains locked on the signal, is established when the carrier-phase data are processed; The initial bias in a carrier-phase observation of an arbitrary number of cycles. The initial phase measurement made when a GPS receiver first locks onto a GPS signal is ambiguous by an integer number of cycles because the receiver has no way of knowing the exact number of carrier wave cycles between the satellite and the receiver.

- **ambiguity resolution**

If the initial integer ambiguity value for each satellite-receiver pair could be determined, then the ambiguous integrated carrier phase measurement can be corrected to create an unambiguous, but very precise (millimetre observation accuracy), receiver-satellite distance measurement. A solution using the corrected carrier phase observations is known as an "ambiguity-fixed" or "bias-fixed" solution. The mathematical process or algorithm for determining the value for the ambiguities is Ambiguity Resolution. Tremendous progress has been made in AR techniques, making today's carrier phase-based GPS systems very efficient by cutting down the length of observation data needed (resulting in so-called "rapid static surveying" techniques) and even allowing this process to occur while the receiver is itself in motion (in so-called "on-the-fly" AR techniques). (In practice, the AR process and the ambiguity-fixed solutions are carried out on the double-differenced carrier phase observables, not on the one-way satellite-receiver measurements.)

- **ami-c (automotive multimedia interface consortium)**

Non-profit corporation of worldwide vehicle manufacturers which aims to develop a set of common specifications for a multimedia interface to motor vehicle electronic systems to accommodate a variety of computer-based electronic devices in the vehicle.

- **amplifier**

A device used to boost the strength of an electronic signal.

- **amplitude**

Height of a radiowave as measured from an imaginary center line to the wave peak.

- **amplitude modulation (am)**

A method of encoding a message on the carrier signal by altering the height of the signal while keeping its frequency constant.

- **amps**

Analog Mobile Phone System. Non-digital cellular mobile phones.

- **analog**

A form of transmitting information characterized by continuously variable quantities, as opposed to digital transmission, which is characterized by discrete bits of information in numerical steps. An analog signal is responsive to changes in light, sound, heat and pressure.

- **analog data**

Data represented by a physical quantity that is considered to be continuously variable and whose magnitude is made directly proportional to the data or to a suitable function of the data.

- **analog mobile phone system (amps)**

Non-digital cellular mobile phones.

- **analog signal**

The principal feature of analog signals is that they are continuous. In contrast, digital signals consist of values measured at discrete intervals.

- **analog-to-digital conversion (adc)**

Process of converting analog signals to a digital representation. DAC represents the reverse translation.

- **anik**

The Canadian domestic satellite system that transmits Canadian Broadcasting Corporation's (CSC) network feeds throughout the country. This system also carries long distance voice and data services throughout Canada as well as some transborder service to the U.S. and Mexico.

- **anomalistic period**

The elapsed time between two successive perigees of a satellite.

- **antarctic circle**

An imaginary circle on the surface of the earth at 66.5°S lat.or 23.5° north of the South Pole. It marks the southernmost point at which the sun can be seen at the winter solstice (about June 22) and the northernmost point of the southern polar regions at which the midnight sun is visible.

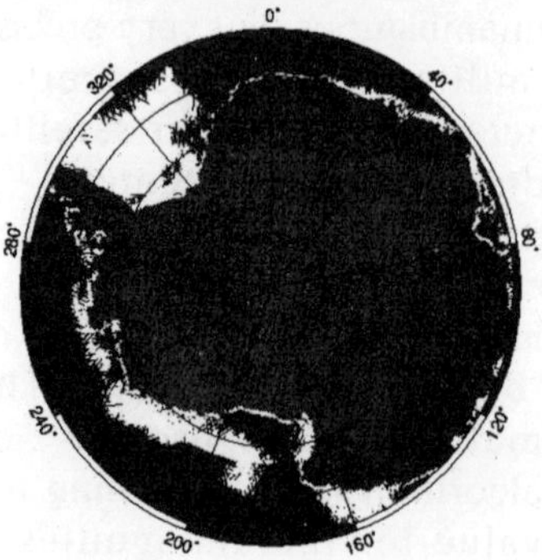

- **antenna**

That part of the GPS receiver hardware which receives (and sometimes amplifies) the incoming L-Band signal. Antennas come in all shapes and sizes, but most these days use so-called "microstrip" or "patch" antenna elements. The geodetic antennas, on the other hand, may use a "choke-ring" to mitigate any multipath signals; Thales Navigation offers a variety of global navigation satellite antennas designed to pick up navigation satellite signals. Antennas range from simpler microstrip devices to complex choke ring antennas that mitigate the effects of multipath scattering sometimes caused when signals are reflected off tall buildings, cliffs, mountains or other localized terrain; A device for transmitting and receiving radio waves. Depending on their use and operating frequency, antennas can take the form of a single piece of wire, a di-pole a grid such as a yagi array, a horn, a helix, a sophisticated parabolic-shaped dish, or a phase array of active electronic elements of virtually any flat or convoluted surface.

- **antenna splitter**

An attachment which can be used to split the antenna signal into two, so that it may be fed to two GPS receivers. Such a configuration forms the basis of a Zero Baseline test.

- **anti spoofing**

AS Encryption of the P-code signal transforming it to Y-code which is unavailable to civilian users; Encryption of the P code signal transforming it to Y code which is unavailable to civilian users. AS prevents an encryption-keyed GPS receiver from being "spoofed" by a bogus, enemy-generated GPS P-code signal.

- **anywhere fix**

The ability of a receiver to start position calculations without being given an approximate location and time.

- **aperture**

A cross sectional area of the antenna which is exposed to the satellite signal; In image scanning, the size of the sensitive spot that moves over the image. In an antenna, an imaginary cylinder in space surrounding the radiating elements.

- **apogee**

The point in an elliptical satellite orbit which is farthest from the surface of the earth. Geosynchronous satellites which maintain circular orbits around the earth are first launched into highly elliptical orbits with apogees of 22,237 miles. When the communication satellite reaches the appropriate apogee, a rocket motor is fired to place the satellite into its permanent circular orbit of 22,237 miles; The point in a satellite's orbit where it is farthest from earth; The point in an elliptical satellite orbit which is farthest from the center of the earth (the opposite of perigee).

- **apogee kick motor (akm)**

Rocket motor fired to circulate orbit and deploy satellite into geostationary orbit.

- **arc-digitized raster graphics (adrg)**

A United States military map format for providing maps on CD.

- **arcgis**

A family of software products that form a complete GIS (Geographic Information System).

- **arcinfo**

Comprehensive software in the arc GIS family that has advanced geoprocessing and data conversion capabilities.

- **arctic circle**

An imaginary circle on the surface of the earth at 66.5°N latitude, or,

23.5° south of the North Pole. It marks the northernmost point at which the sun can be seen at the winter solstice (about Dec. 22) and the southernmost point of the northern polar regions at which the midnight sun is visible.

- **argos**
  Advanced Research and Global Observation Satellite. An advanced research and development satellite launched by the United States Air Force. Carries various space and atmospheric experiments.

- **argument of perigee**
  The polar angle that locates the perigee point of a satellite in the orbital plane; drawn between the ascending node, geocenter, and perigee; and measured from the ascending node in the direction of satellite motion.

- **arof**
  Ambiguity Resolution on the Fly Fast ambiguity resolution GPS surveying where the solution is determined while the receiver is in motion.

- **artifact**
  Any anomaly, distortion or unnatural component in an image — usually associated with digital video transmission.

- **ascending node**
  The point at whi[illegible] object passes the reference pla[illegible] from south to north; The point on the ground track of a satellite orbit where the sub-satellite point(SSP) crosses the equator from the Southern Hemisphere into the Northern Hemisphere.

- **ascii (american standard code for information interchange)**
  A standardized format for computer system text coding.

- **asf**
  Additional Secondary Factor. Error propagation time caused by radiowaves slowing down as they pass over different terrain. This is in addition to slowing down as radiowaves pass over sea water.

- **asic**
  Application Specific Integrated Circuit

- **aspect ratio**
  The ratio of the width to the height of an image. For example, analog television uses a 4:3 (1.33) aspect ratio, while HDTV has an aspect ratio of 16:9 (1.78).

- **asset monitoring**
  tracking the location and usage of business equipment such as ships, measuring equipment, etc.

- **asynchronous transfer mode (atm)**
  A method of transmitting/switching video/data using very high speed and high bandwidth.

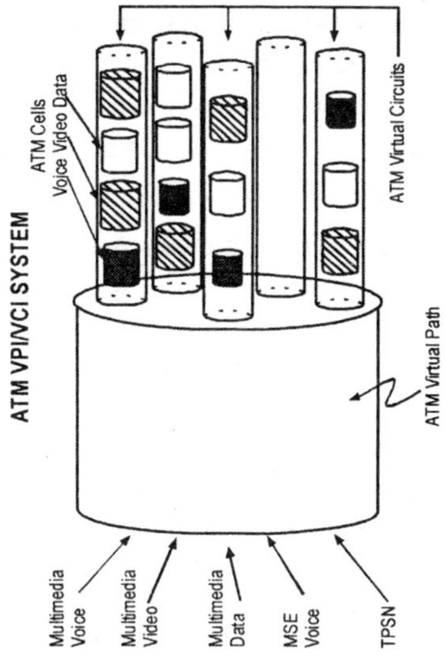

- **asynchronous transmission**

A method of data transmission which allows characters to be sent at irregular intervals by preceding each character with a start bit, and following it with a stop bit.

- **atlas florae europaeae (afe)**

A modified from the Military Grid Reference System (MGRS).

- **atomic clock**

A very precise clock that operates using the elements cesium or rubidium. A cesium clock has an error of one second per million years. GPS satellites contain multiple cesium and rubidium clocks; An atomic clock is an extremely accurate timekeeping device regulated by the natural regular oscillations of an atom or molecule. An atomic clock powered by a hydrogen atom (maser) is accurate to 1 part in 2 quadrillion. A cesium atom powered clock has an error of about one second in one million years. GPS satellites carry multiple atomic clocks, regulated by these precise atomic vibrations, to ensure accuracy. The GPS Master Control Station uses cesium clocks and a hydrogen maser clock.

- **atsc (advanced television systems committee)**

An international organization of approximately 200 members that is establishing voluntary technical standards for advanced television systems. ATSC Digital TV Standards include digital HDTV (High Definition Television), SDTV (Standard Definition Television), data broadcasting, multichannel surround-sound audio, and Satellite direct-to-home broadcasting.

- **attenuation**

A reduction in signal strength; The loss in power of electromagnetic signals between transmission and reception points.

- **attitude control**

The orientation of the satellite in relationship to the earth and the sun.

- **attribute**

a characteristic which describes a Feature. Attributes can be thought of as questions which are asked about the Feature; A characteristic which describes a Feature. Attributes can be thought of as questions which are asked about the Feature. Typically associated with geospatial data gathering for inclusion within Geographic Information Systems.

- **audio subcarrier**

  The carrier between 5 MHz and 8 MHz containing audio (or voice) information inside of a video carrier.

- **autolocate®**

  This is a proprietary feature of Garmin GPS receivers. A Garmin unit displays the "AutoLocate" status when it is looking for and collecting data from satellites that were visible at its' last known or initialized position (almanac data), but it has not collected enough data to calculate a position fix.

- **automatic frequency control (afc)**

  A circuit which automatically controls the frequency of a signal.

- **automatic gain control (agc)**

  A circuit which automatically controls the gain of an amplifier so that the output signal level is virtually constant for varying input signal levels.

- **automatic packet reporting system (aprs) terminal information service (atis)**

  A protocol that uses packet radio to transmit data that depicts the position of a station with other.

- **automatic terminal information service (atis)**

  It is a continuous broadcast of recorded information at selected high activity airports.

- **automatic vehicle location**

  A system of providing real-time location information for emergency vehicles, delivery trucks, service vehicles, etc; The technique of using a navigation system, such as GPS, to determine a vehicle's position; The technique of using a navigation system to determine a vehicle's position, which is then transmitted to a central unit that monitors and tracks the vehicle's position and movement.

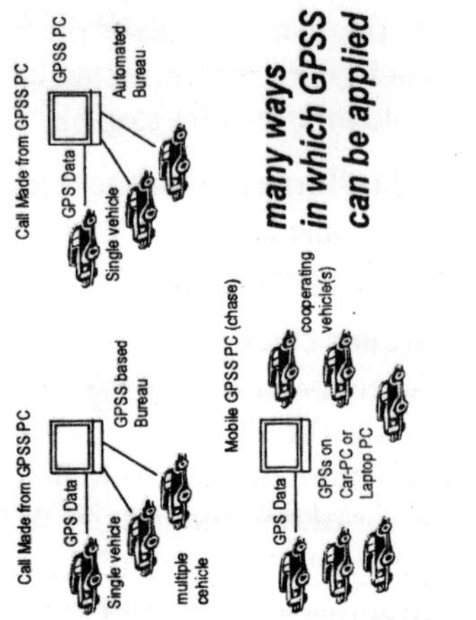

- **automatic vehicle monitoring (avm)**

  The technique of using a navigation system to determine a vehicle's position, which is then transmitted to a central unit that monitors and tracks the vehicle's position and movement.

- **autonomous underwater vehicle**

an unmanned submersible launched from a 'mother-ship' but not connected to it via a cable. Propulsion and control are autonomous and use pre-defined mission protocols.

- **availability**

The number of hours per day that a particular location has sufficient satellites (above the specified elevation angle and less than the specified PDOP value) to make a GPS position fix; The ratio, expressed as a percentage, of the total time a functional unit or service is capable of being used or is available to be used during a given interval to the length of the interval; e.g., if the unit is not capable of being used for 20 minutes in a week, the availability is 99.8 percent (10080 - 20 minutes/ 10080 minutes X 100).

- **avi (automated vehicle identification)**

A system which transmits signals from an on-board tag or transponder to roadside receivers for uses such as electronic toll collection (ETC) and stolen vehicle recovery.

- **az/el mount**

Antenna mount that requires two separate adjustments to move from one satellite to another.

- **azimuth**

The horizontal direction from one point on the earth to another, measured clockwise in degrees (0-360) from a north or south reference line. An azimuth is also called a bearing; Horizontal direction of a celestial point from a terrestrial point, expressed as the angular distance from 0 degrees?(reference) clockwise through 360 degrees. The reference point is generally True North, but may be Magnetic North, or relative to a fixed point on a vehicle; This is the direction of a celestial object, measured clockwise around the observer's horizon from north. So an object due north has an azimuth of 0°, one due east 90°, south 180° and west 270°. Azimuth and altitude are usually used together to give the direction of an object in the topocentric coordinate system.
The definition from Webster's New Collegiate Dictionary states, "an arc of the horizon measured between a fixed point such as true north and the vertical circle passing through the center of an object." Azimuth is an important consideration in locating a satellite for transmission or reception of RF signals. The azimuth expressed in degrees of a circle will be the horizontal angle of rotation that the ground antenna must be rotated though to point at the specific satellite. Azimuth angles

for any satellite may be calculated given the latitude and longitude of the ground station and the location of the satellite in geosynchronous orbit relative to true north.

- **azimuth ring**
  The dial on a compass, marked with zero to 360 degree markings.

- **backhaul**
  A terrestrial communications channel linking an earth station to a local switching network or population center; A process where a television signal is transmitted from a remote site to a central site where the backhaul signal is integrated into a program which is being broadcast from the central site.

- **backlight**
  Selectable on/off illumination that lights both the display and keypad on Lowrance units for enhanced screen viewing when fishing at night and in low light.

- **backoff**
  The process of reducing the input and output power levels of a traveling wave tube to obtain more linear operation.

- **backtrack**
  Navigation of a route in reverse order from the last position fix in succession to the first. Reverse of navigating a route in normal sequence.

- **band pass filter**
  An active or passive circuit which allows signals within the desired frequency band to pass through but impedes signals outside this pass band from getting through.

- **bandwidth**
  A measure of the information-carrying capacity of a signal, expressed as the width of the spectrum of that signal (frequency domain representation) in Hertz; A measure of spectrum (frequency) use or capacity. For instance, a voice transmission by telephone requires a bandwidth of about 3000 cycles per second (3KHz). A TV channel occupies a bandwidth of 6 million cycles per second (6 MHz) in terrestrial Systems. In satellite based systems a larger bandwidth of 17.5 to 72 MHz is used to spread or "dither" the television signal in order to prevent interference; The capacity of a telecom line to carry signals. The necessary bandwidth is the amount of spectrum required to transmit the signal without distortion or loss of information. FCC rules require suppression of the signal outside the band to prevent interference.

- **bas (brake assist system)**
  Safety system amplifying the brake force in case of emergency.

- **base station**

A receiver that is set up on a known location specifically to collect data for differentially correcting rover files. The base station calculates the error for each satellite and, through differential correction, improves the accuracy of GPS positions collected at unknown locations by a roving GPS receiver; Also called a Reference Station. In GPS navigation, this is a receiver that is set up on a known location specifically to collect data for differentially correcting data files of another receiver (which may be referred to as the "mobile" or "rover" receiver). In the case of pseudo-range-based Differential GPS (DGPS) the base station calculates the error for each satellite and, through differential correction, improves the accuracy of GPS positions collected at unknown locations by another (roving) GPS receiver. For GPS Surveying techniques, the receiver data from the base station is combined with the data from the other receiver to form double-differenced observations, from which the baseline vector is determined; The basic direct output signal in an intermediate frequency based obtained directly from a television camera, satellite television receiver, or video tape recorder. Baseband signals can be viewed only on studio monitors. To display the baseband signal on a conventional television set a "modulator" is required to convert the baseband signal to one of the VHF or UHF television channels which the television set can be tuned to receive; The bandwidth of the Modulating (message) signal. A transmission medium through which digital signals are sent without frequency shifting. In general, only one communication channel is available at any given time. Ethernet is an example of a baseband network.

- **baseline**

A baseline consists of a pair of stations for which simultaneous GPS data has been collected; Mathematically expressed as a vector of coordinate differences between the two stations, or an expression of the coordinates of one station with respect to the other (whose coordinates are assumed known, and is typically referred to as a "Base" or "Reference" Station); Vector the difference in three-dimensional co-ordinates (X, Y, Z) computed from the difference in simultaneous carrier phase observations at two or more receivers.

- **basemap**

  Garmin mapping units come with permanently built-in basemaps, which typically include coverage of oceans, rivers, and lakes; principal cities, smaller cities, and towns; interstates, highways, and local thoroughfares; and railroads, airports, and political boundaries. Basemaps are available in a variety of global coverage areas, depending on the user's needs.

- **baud**

  The rate of data transmission based on the number of signal elements or symbols transmitted per second. Today most digital signals are characterized in bits per second; A description of the rate of data transmission usually expressed in bit rate measurements of thousand bits per second (Kb/s). A 14.4 Baud modem for example will accommodate a bit rate of 14,400 bits per second.

- **beacon**

  Stationary transmitter that emits signals in all directions (also called a non-directional beacon). In DGPS, the beacon transmitter also broadcasts pseudorange correction data to nearby GPS receivers for greater accuracy.

- **beamwidth**

  The angle or conical shape of the beam the antenna projects. Large antennas have narrower beamwidths and can pinpoint satellites in space or dense traffic areas on the earth more precisely. Tighter beamwidths thus deliver higher levels of power and thus greater communications performance; The angular width of the main lobe of an antenna radiation pattern as measured between predetermined points on the lobe, usually the half-power (-3 dB) points.

- **bearing**

  The compass direction from a position to a destination, measured to the nearest degree (also call an azimuth). In a GPS receiver, bearing usually refers to the direction to a waypoint; Also referred to as the Azimuth. The compass direction from a position to a destination. The "north" direction is "zero bearing", and the angle is measured clockwise through 360°. May be referred to a number of "north" directions, including magnetic north, (projection) grid north, or geographic north; For G.P.S. navigation, it generally refers to the compass reading to navigate from one position to another measured to the nearest degree. In traditional nautical navigation, it is referenced a differently in degrees from quadrants of the compass. See G.P.S. for Land Navigation by Michael Ferguson for more details on this topic.

- **beat frequency**

Either of two additional frequencies generated when two frequencies are mixed. One beat frequency is equal to the sum of the frequencies, the other the difference of the frequencies.

- **bench mark**

A relatively permanent material object, natural or man-made, with a known elevation. A bench mark can be used as a reference point when navigating a route or in determining the elevation of nearby land features.

- **bias**

All GPS measurements are affected by biases and errors. Their combined magnitudes will affect the accuracy of the positioning results (they will bias the position or baseline solution). Biases may be defined as being those systematic errors that cause the true measurements to be different from observed measurements by a "constant, predictable or systematic amount", such as, for example, all distances being measured too short, or too long. Biases must somehow be accounted for in the measurement model used for data processing if high accuracy is sought. There are several sources of biases with varying characteristics, such as magnitude, periodicity, satellite or receiver dependency, etc. Biases may have physical bases, such as the atmosphere effects on signal propagation or ambiguities in the carrier phase measurements, but may also enter at the data processing stage through imperfect knowledge of constants, for example any "fixed" parameters such as the satellite ephemeris information, station coordinates, velocity of light, antenna height errors, etc. Random errors will not bias a solution. However, outlier measurements, or measurements significantly affected by multipath disturbance (which may be considered a transient, unmodelled bias), will bias a solution if the proportion of affected measurements is relatively high compared to the number of unaffected measurements. For this reason, long period static GPS Surveying is more accurate (less likely to be biased) than "rapid static surveying" or kinematic (single-epoch) positioning.

- **binary biphase modulation**

The phase modulation technique used to transmit a GPS signal. The phase of a carrier generated by a satellite is shifted by 180 degrees when there is a code or message binary signal level transition, either from 0 to 1 (normal to mirror image) or from 1 to 0 (mirror image to normal).

- **binary phase shift keying (bpsk)**
  A digital modulation scheme used in transmission communications.

- **binary pulse code modulation**
  Pulse modulation using a string of binary numbers (codes). This coding is usually represented by ones and zeros with definite meanings assigned to them, such as changes in phase or direction of a wave.

- **binary shift-key (bsk) modulation**
  BSK is a modulation technique by which a binary message, such the Navigation Message or the PRN codes (consisting of 0's and 1's), is imprinted on the carrier wave. Unlike Amplitude Modulation (AM) and Frequency Modulation (FM), BSK Modulation does not alter the signal level (the "amplitude") or the carrier wavelength (the "frequency"). At a change in value of the message from 0 or 1, or from 1 to 0, the carrier wave is reversed (the phase is "flipped" by 180°). All reversals take place at the zero-crossings of the carrier (sine) wave (i.e., where the phase is zero).

- **bipm**
  Bureau International des Poids et Measure.The international bureau of weights and measurements, in Paris. Responsible for the UTC time scale.

- **bipolar biphase shift key (bpsk)**
  BPSK is the modulation technique for carrier phase encoding on global navigation satellites using binary bit transition that results in a 180-degree shift of the carrier.

- **bird**
  Slang for a communications satellite located in geo-synchronous orbit.

- **b-isdn (broadband isdn)**
  An ISDN (Integrated Services Digital Network) offering broadband capabilities. B-ISDN is a CCITT-proposed service that may include:
  (a) interfaces operating at data rates from 150 to 600 Mb/s,
  (b) use ATM (Asynchronous Transfer Mode) to carry all services over a single, integrated, high-speed packet-switched network,
  (c) have LAN interconnection capability,
  (d) provide access to a remote, shared disk server,
  (e) provide voice/video/data teleconferencing,
  (f) provide transport for programming services such as cable TV,
  (g) provide single-user controlled access to remote video sources,
  (h) handle voice/video telephone calls and
  ](i) access shop-at-home and other information services.

- **bit**

(Abbreviation for "binary digit") The smallest increment of digital information. Often referred to as a 1 or 0 in the binary system. Also referred to as an 'on' or 'off' when referring to the mechanical operation of bits in the computer.

- **bit error rate**

The fraction of a sequence of message bits that are in error. A bit error rate of 10-6 means that there is an average of one error per million bits.

- **bit rate**

The speed of a digital transmission, measured in bits per second; The amount of data being transported, measured relative to quantity over time in bits per second (thousand bits per second or Kb/s, million bits per second or Mb/s, billion bits per second or Gb/s and trillion bits per second or Tb/s). Summary of bitrate notations:

Bit: 10p0 (1 bit)

Kilobit: 10p3 (1,000 bits)

Megabit: 10p6 (1,000,000 bits)

Gigibit: 10p9 (1,000,000,000 bits)

Terabit: 10p12 (1,000,000,000,000 bits)

Petabit: 10p15 (1,000,000,000,000,000 bits)

Exabit: 10p18 (1,000,000,000,000,000,000 bits)

Zettabit: 10p21 (1,000,000,000,000,000,000,000 bits)

Yottabit: 10p24 (1,000,000,000,000,000,000,000,000 bits)

A 40 Exabit file would contain all of the words ever spoken by human beings.

- **bitmap image**

An image with 1 bit of color information per pixel, also known as a bitmapped image.

- **blanking**

An ordinary television signal consists of 30 separate still pictures or frames sent every second. They occur so rapidly, the human eye blurs them together to form an illusion of moving pictures. This is the basis for television and motion picture systems. The blanking interval is that portion of the television signal which occurs after one picture frame is sent and before the next one is transmitted. During this period of time special data signals can be sent which will not be picked up on an ordinary television receiver.

- **block down converter**

A device used to convert the 3.7 to 4.2 KHz signal down to UHF or lower frequencies (1 GHz and lower).

- **blue book**

Term referring to the Federal Geodetic Control Subcommittee

publication "Input Formats and Specifications of the National Geodetic Survey Data Base" which defines the file formats for data submission to the National Geodetic Reference System; Slang term derived from a blue NGS reference book. The book contains information and formats required by NGS for survey data that is submitted to be considered for use in the national network.

- **bluetooth**

  A standard of radio technology that makes it possible to transmit signals over short distances between telephones, computers and other devices.

- **b-mac**

  A method of transmitting and scrambling television signals. In such transmissions MAC (Multiplexed Analog Component) signals are time-multiplexed with a digital burst containing digitized sound, video synchronizing, authorization, and information.

- **bnc**

  Bayonet N Connector. A common type of antenna connector. Push and Turn to lock

- **brg bearing**

  The compass direction from your position to a destination, measured to the nearest degree.

- **broad beam**

  A single large circular beam that covers a large geographic area.

- **broadband**

  Broadband is a descriptive term for evolving digital technologies that provide consumers a signal switched facility offering integrated access to voice, high-speed data service, video-demand services, and interactive delivery services.

- **broadcast**

  The sending of one transmission to multiple users in a defined group; Using the electronic media to reach a wide-area audience.

- **bss (broadcast satellite service)**

  The segment of Ku-band satellite service established by the FCC which is provided by high power satellites (or clusters of satellites) which must be separated in orbit by 9 degrees. The frequencies of RF signals transmitted to and received from the BSS satellites are higher than those permitted for FSS satellites. The BSS transmission frequency from high-power satellites to satellite receivers has been established by the FCC between 12.2 - 12.7 GHz.

- **buoy**

  A navigation aid (usually found in marine charts) that is floating in water. The position can vary due to the magnitude and direction of the current.

- **business television**

Corporate communications tool involving video transmissions of information via satellite. Common uses of business television are for meetings, product introductions and training.

- **byte**

(Abbreviation for "binary term The simplest combination of bits producing recognizable information such as a number or word. For example, the number 50 is represented by a byte made up of the eight bits 0,0,1,1,0,0,1 and 0.

- **c band**

This is the band between 4 and 8 GHz with the 6 and 4 GHz band being used for satellite communications. Specifically, the 3.7 to 4.2 GHz satellite communication band is used as the down link frequencies in tandem with the 5.925 to 6.425 GHz band that serves as the uplink. A portion of the Radio Frequency (RF) spectrum located between 4 GHz and 8 GHz, a part of which is dedicated to satellite communications. Satellite downlink frequencies are located between 3.7 GHz and 4.2 GHz and uplink frequencies are located between 5.925 GHz and 6.425 GHz; A band ot trequencies used for satellite and terrestrial communications. The range of frequencies from 4 to 6 gigaHertz (billion cycles per second) is used by most communications satellites.

- **c code**

The less accurate Civilian GPS Signal. See Selective Availability.

- **c/a (clear acquisition) code**

A sequence of 1023 bits that repeat every millisecond. Each GPS satellite broadcasts a unique 1023-bit code to enable receivers to distinguish between satellites. The C/A-Code is only on the L1 carrier frequency on GPS satellites.

- **c/a (coarse acquisition) code**

Spread spectrum direct sequence code that is used primarily by commercial GPS receivers to determine the range to GPS satellite. The code is a sequence of 1023 pseudorandom binary biphase modulations at a chipping rate of 1.023 MHz; The standard (Course/Acquisition) GPS code. A sequence of 1023 pseudo-random, binary, biphase modulations on the GPS carrier at a chip rate of 1.023 MHz. Also known as the "civilian code."; Coarse Acquisition code-The GPS signal for civilian use (also used by the military to get an initial fix).

- **c/no**

Carrier-to-noise ratio measured either at the Radio Frequency (RF) or Intermediate Frequency (IF)

- **c/t**

Carrier-to-noise-temperature ratio.

- **cache**

A small fast memory which holds recently accessed data, designed to speed up subsequent access to the same data. Most often applied to processor-memory access but also used for a local copy of data accessible over a network, etc.

- **cadastral survey**

Cadastral survey is a land survey that defines boundaries, property lines, and other measurements pertaining to an official register of ownership, known as a cadastre. A cadastral survey helps determine the expanse and value of real property for resale and taxation.

- **cadastre**

A public record, survey, or map of the value , extent, and ownership of land as a basis of taxation.

- **call**

(1) Any demand to set up a connection.

(2) A unit of traffic measurement.

- **call sign**

Each amateur radio station is assigned a call sign by the government of that country which allows the operator to transmit in the amateur radio bands.

- **can (controlled area network)**

Data bus which enables computerised exchanges in a high bandwidth.

- **carrier**

A radio wave having at least one characteristic (e.g., frequency, amplitude, phase) that can be varied from a known reference value by modulation. In the case of GPS there are two transmitted carrier waves:

(a) L1 at 1575.42MHz,

(b) L2 at 1227.60MHz, modulated by the Navigation Message (both L1 and L2), the P-Code (both L1 and L2) and the C/A-Code (L1).

A radio wave that conveys or carries some kind of modulation; The basic radio, television, or telephony center of frequency transmit signal. The carrier in an analog signal. is modulated by manipulating its amplitude (making it louder or softer) or its frequency (shifting it up or down) in relation to the incoming signal. Satellite carriers operating in the analog mode are usually frequency modulated.

- **carrier beat phase**

The phase of the signal that remains after the incoming satellite carrier signal is mixed with the constant reference frequency in the receiver. This is used to determine Doppler shift of the satellite's signal.

- **carrier frequency**

The frequency of an unmodulated output of a radio transmitter. The

GPS L1 carrier frequency is 1575.42 MHz; Frequency of the unmodulated output of a radio transmitter. 1. L1 carrier signal broadcasts at 1575.42 MHz (19 cm wavelength). 2. L2 carrier signal broadcasts at 1227.60 MHz (24 cm wavelength).

The basic frequency of an unmodulated radio signal is known as the carrier frequency. GPS satellite navigation signals are broadcast on two L-band frequencies, L1 and L2. L1 broadcasts at 1575.42 Mhz, while L2 broadcasts at 1227.6 Mhz.

- **carrier phase**

GPS measurements made on the L1 or L2 carrier signal. May refer to the fractional part of the L1 or L2 carrier wavelength (approximately 19cm for L1, 24cm for L2), expressed in units of metre, cycles, fraction of a wavelength or angle. (One cycle of L1 is equivalent to one wavelength, and similarly for L2.) In carrier phase-based positioning, such as employed in GPS Surveying techniques, carrier phase may also refer to the accumulated or integrated measurement which consists of the fractional part plus the whole number of wavelengths (or cycles) since signal lock-on; Carrier phase is the fraction of a cycle often expressed in degrees, where 360 degrees equals a complete cycle. Carrier phase can also mean the number of complete cycles plus a fractional cycle. A survey-grade GPS receiver can lock-on to a satellite and count the number of whole cycles of the carrier frequency, measuring the cumulative phase of the signal. This is often referred to as integrated Doppler.

- **carrier phase gps**

GPS measurements based on the L1 or L2 carrier signal.

- **carrier phase tracking**

This is an accurate and sophisticated method of determining position requiring two or more special receivers that measure small differences in the radio signals using the radio carrier frequency as a standard.

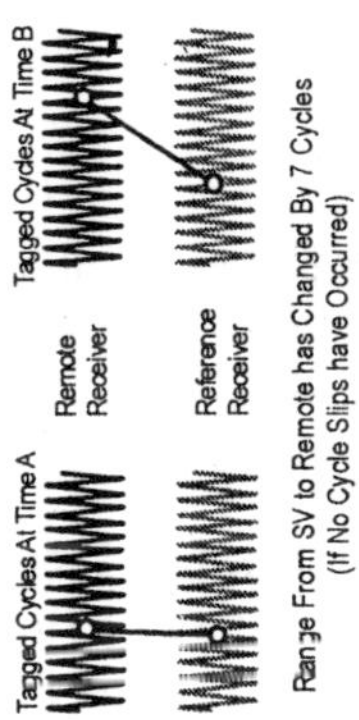

- **carrier to noise ratio (c/n)**

The ratio of the received carrier power and the noise power in a given bandwidth, expressed in dB. This figure is directly related to G/T and S/N; and in a video signal the higher the C/N, the better the received picture.

- **carrier-aided tracking**
  A signal processing strategy that uses the GPS carrier signal to achieve an exact lock on the pseudo random code. More accurate than standard approach; A technique to improve accuracy by using the GPS carrier signal to get a more exact lock on the pseudorandom code; A signal processing technique that uses the GPS carrier signal to achieve an exact lock on the pseudo random code generated by the GPS satellite. Carrier-aided tracking is more accurate than standard C/A Code tracking.

- **carrier-to-noise power density (c/n0)**
  The ratio of the power level of a signal carrier to the noise power in a 1-Hz bandwidth. This is a key parameter in the analysis of GPS receiver performance. Nominal GPS receiver C/N0 values often are in the 40 to 50-dB-Hz range.

- **carrier-tracking loop**
  A module in a GPS receiver that demodulates, or extracts, the satellite message by aligning the phase of the receiver's local oscillator signal with the phase of the frequency-shifted, received carrier. Once the local oscillator signal is locked to the carrier, its phase can be measured to provide the carrier-phase observation.

- **cartography**
  The art or technique of making maps or charts. Many GPS receivers have detailed mapping- or cartography-capabilities.

- **cassegrain antenna**
  The antenna principle that utilizes a subreflector at the focal point which reflects energy to or from a feed located at the apex of the main reflector.

- **cavitation**
  The sudden formation and collapse of air bubbles in water due to turbulence. If air bubbles pass over the face of the transducer, the signal from the transducer is reflected back by the air bubbles. These reflections are very strong and will result in noise which will interfere with bottom, structure and fish signals, making them difficult or impossible to see.

- **cdma**
  Code division multiple access. Refers to a multiple-access scheme where stations use spread-spectrum modulations and orthogonal codes to avoid interfering with one another.

- **cellular technology**
  This term, often used for all wireless phones regardless of the technology they use, derives from cellular base stations that receive and transmit calls. Both cellular and PCS phones use cellular technology.

• cep

Circular Error Probable. The radius of a circle within which fifty per cent of positioning solutions fall. CEP is used to achieve horizontal accuracy.

• cf

Compact Flash-A standard that PDAs, laptops, and other computer-related devices use to accept removable accessories such as additional memory or GPS receivers.

• changeable message sign (cms) - dynamic message sign (dms) - variable message sign (vms)

Electronic sign capable of displaying real-time traffic information.

• channel

In a receiver, a path for an electronic signal that is reserved for a specific satellite and used for various functions; A channel of a GPS receiver consists of the circuitry necessary to receive the signal from a single GPS satellite; Circuitry necessary to receive the signal from one GPS satellite. Channel refers to a set of hardware in a receiver that detects, locks on and continuously tracks the signal from a single navigation satellite. The more receiver channels available, the greater number of satellite signals a receiver can simultaneously lock-on and track; A frequency band in which a specific broadcast signal is transmitted. Channel frequencies are specified in the United States by the Federal Communications Commission. Television signals require a 6 MHz frequency band to carry all the necessary picture detail.

• chart speed

The chart is the picture of the underwater world you see on the display, and the chart speed is how fast that picture moves...or scrolls...across the screen from right to left. This scroll speed is adjustable to help display better fish arches...or to suit your viewing preferences when using fish symbols.

• chartplotter

A device that overlays GPS data on to marine navigational charts.

• chip

Binary elements or digits that unlike bits, convey no information. A PRN code consists of a sequence of chips; The transition time for individual bits in the pseudo-random sequence. Also, an integrated circuit. Also a snack food. Also a betting marker.

• chip rate

The number of chips sent per second; Number of chips per second (e.g., C/A code = 1.023 MHz).

• chrominance

In color video systems, the signal component(s) that describe color difference information.

- **cib**

  Controlled image base.

- **circuit mode**

  A circuit switched operational mode for transferring (transporting and switching) user information through a network. Contrast with packet switching mode.

- **circular error probable (cep)**

  A statistical measure of the horizontal precision. The CEP value is defined as a circle of a specified radii that encloses 50 per cent of the data points. Thus, half the data points are within a 2D CEP circle and half are outside the circle.

- **circular polarization**

  Unlike many domestic satellites which utilize vertical or horizontal polarization, the international Intelsat satellites transmit their signals in a rotating corkscrew-like pattern as they are down-linked to earth. On some satellites, both right-hand rotating and left-hand rotating signals can be transmitted simultaneously on the same frequency; thereby doubling the capacity of the satellite to carry communications channels; A special method of radio energy emission where the electric and magnetic field vectors rotate about the central axis of radiation. As viewed along the radiation path, the rotation directions are considered to be right-hand-circular(RHCP) if the rotation is clockwise and left-hand-circular(LHCP) if the rotation is counterclockwise.

- **cits**

  Close Air Support Integrated Targeting System is a program developed by Northport Systems Inc. for integrating the Global Positioning System with laser target locators.

- **clamp**

  A video processing circuit that removes the energy dispersal signal component from the video waveform.

- **clarke 1866**

  The reference ellipsoid for the NAD27 coordinate system. a = 6378206.400 m, b = 6356583.800 m, 1/f = 294.978 698 200

- **clarke orbit**

  That circular orbit in space 22,237 miles from the surface of the earth at which geosynchronous satellites are placed. This orbit was first postulated by the science fiction writer Arthur C. Clarke in Wireless World magazine in 1945. Satellites placed in these orbits, although traveling around the earth at thousands of miles an hour, appear to be stationary when viewed from a point on the earth, since the earth is rotating upon its axis at the same angular rate that the satellite is traveling around the earth.

- **class of survey**
Class of Survey is a means of categorising the internal quality, or precision of a survey. The number of categories, the notation applied, and the accuracy tolerances defining the transition from one class to another are defined by individual nations. Typically they are based on traditional geodetic surveying categories, supplemented by several extra categories of higher precision applicable to GPS Surveying and GPS Geodesy techniques, and may be different for horizontal surveys and vertical surveys. The attachment of a particular Class "label" (e.g. A, B, etc.) to a survey, comprising a few or many points within a "network", carried out using GPS or any other technique, is performed as part of the process of "network adjustment" in which the relative error ellipses (in the horizontal case) between coordinated stations are computed and compared with the accuracy standards that must be met for various categories of Class.

- **clear acquisition (c/a) code**
C/A-Code consists of a sequence of 1023 bits (0 or 1) that repeats every millisecond. Each satellite broadcasts a unique 1023-bit sequence that allows a receiver to distinguish between various satellites. The C/A-Code modulates only the L1 carrier frequency on GPS satellites. The C/A-Code allows a receiver to quickly lock-on to a satellite.

- **cliff effect**
A characteristic of the digital transmission of RF signals where there is a radical change in reception quality which results from a small change in reception power. By comparison, when an analog RF signal approaches the fringes of acceptable reception power, the television picture begins to experience gradual degradation with increasing sparkles or snow. As a digital RF signal reaches the fringes of acceptable reception power, there is no discernible degradation of picture quality until the level of reduced power reaches a threshold. At that point, picture quality changes from perfect to no picture.

- **clinometer**
A device, usually similar to a compass, used to measure vertical angles, as in the slope of a hill.

- **clipper chip**
An asymmetric encryption chip proposed for use by the United States government so that the pervasive encryption of communication channels shall not protect both the law-abiding and law-breakers. Designated MYK-78, a secret key is created at the time of manufacture and placed in escrow. With court approval,

government agencies can obtain this key and employ it to decipher encrypted messages obtained through wiretaps.

- **clock**
  Your local time.

- **clock aiding**
  An accuracy enhancement technique in which an additional atomic clock (rubidium) provides accurate time to the receiver for calculating satellite clock frequency, phase bias, and clock drift.

- **clock bias**
  The difference between the receiver or satellite clock's indicated time and a well-defined time scale reference such as UTC (Coordinated Universal Time), TAI (International Atomic Time) or GPST (GPS Time).

- **clock offset**
  A constant difference in the time reading between two clocks, normally used to indicate a difference between two time zones.

- **closed captioning**
  A service for persons with hearing disabilities that translates television program dialog into written words on the television screen.

- **closed circuit television (cctv) - surveillance cameras**
  TxDOT installed and maintained cameras located at approximately one (1) mile spacing along the freeway to locate or verify incidents. This is a tool that provides a quicker response time for emergency personnel. (i.e., police, fire, and medical).

- **cmg**
  Course Made Good. How you are progressing towards your next waypoint.

- **coarse acquisition**
  Code a spread spectrum direct sequence code that is used primarily by commercial GPS receivers to determine the range to the transmitting GPS satellite.

- **coarse acquisition code**
  A family of PRN codes transmitted by GPS satellites. Each satellite is assigned one of 32 unique codes in the family. Each code consists of 1,023 chips and is sent at a rate of 1.023 megabits per second. The code sequence repeats every millisecond. The C/A-codes are Gold codes -- PRN codes that are distinguished by a very low cross correlation between any two codes (that is, they are nearly orthogonal). C/A-codes currently are transmitted only on the L1 frequency.

- **coastal confluence zone**
  That area extending from a harbor's entrance 50 nautical miles seaward or to the edge of the continental shelf (100 fathom curve) whichever is greater.

- **code division multile access (cdma)**
  A technique of multiplexing, also called spread spectrum, in which analog signals are converted into digital form for transmission. For each communication channel, the signals are encoded in a sequence known to the transmitter and the receiver for that channel.

- **code phase**
  GPS measurements based on the C/A code; GPS measurements based on the C/A-Code. The term is sometimes restricted to the C/A- or P-Code pseudo-range measurement when expressed in units of cycles; GPS measurements based on pseudo random code (C/A or P) as opposed to carrier of that code.

- **codec**
  Coder/decoder system for digital transmission.

- **code-tracking loop**
  A module in a GPS receiver used to align a PRN code sequence present in a signal coming from a satellite with an identical PRN code sequence generated within the receiver. Alignment is achieved by appropriately shifting the receiver-generated code chips in time so that a particular chip in the sequence is generated at the same instant its twin arrives.

- **cog**
  Course Over Ground, Your current direction of travel relative to a ground position (same as Track).

- **cog course over ground**
  The direction the receiver is moving. COG can be reported in true or magnetic north values.

- **cold start**
  The power-on sequence where the GPS receiver downloads almanac data before establishing a position fix; The ability of a GPS receiver to start providing position updates without the assistance of any almanac information stored in its memory; Used to describe state where a GPS receiver is powered on and does not have valid almanac data available. This lengthens the time required to locate satellites and arrive at the first position solution.

- **collar**
  The area surrounding the actual map itself containing necessary information such as scale, legend, latitude and longitude, etc.

- **co-location**
  Ability of multiple satellites to share the same approximate geostationary orbital assignment frequently due to the fact that different frequency bands are used.

- **color difference**
  In a color video system, signals that represent the difference between

specified colors and the luminance component. They have the property that they go to zero for monochrome images.

- **color subcarrier**
  A subcarrier that is added to the main video signal to convey the color information. In NTSC systems, the color subcarrier is centered on a frequency of 3.579545 MHz, referenced to the main video carrier.

- **colorline™**
  COLORLINE™ helps separate fish and important structures on or near the bottom from the actual bottom. Since more active-feeding fish hold close to hard bottoms and structures, COLORLINE™ helps you find more potentially productive water quickly. By comparing color and width you can distinguish between a hard and soft bottom.

- **commercial leased access**
  Manner through which independent video producers can access cable capacity for a fee.

- **common carrier**
  Any organization which operates communications circuits used by other people. Common carriers include the telephone companies as well as the owners of the communications satellites, RCA, Comsat, Direct Net Telecommunications, AT&T and others. Common carriers are required to file fixed tariffs for specific services; An entity which provides communication transmission facilities for use by other entities or which carries other entity's communications signals. Examples of common carriers include the telephone or telegraph companies and the companies which own communications satellites. Common carriers are subject to tariff regulation and must file rates for specific services with appropriate regulatory agencies such as the FCC or state regulatory agencies. As a rule, common carriers are not permitted to control content.

- **communications assistant**
  A person who facilitates telephone conversation between text telephone users, users of sign language or individuals with speech disabilities through a Telecommunications Relay Service (TRS). This service allows a person with hearing or speech disabilities to communicate with anyone else via telephone at no additional cost.

- **community antenna television (catv)**
  A service through which subscribers pay to have local television stations and additional programs brought into their homes from an antenna via a coaxial cable.

- **companding**
A noise-reduction technique that applies single compression at the transmitter and complementary expansion at the receiver.

- **composite baseband**
The unclamped and unfiltered output of the satellite receiver's demodulator circuit, containing the video information as well as all transmitted subcarriers.

- **compressed video**
A television signal that is transmitted at a slower frame rate than a live video signal.

- **compression**
A technique to reduce the amount of data to be transmitted that reduces the amount of bandwidth needed to transmit video or audio, thus increasing the capacity of a satellite transponder.

- **constellation**
Refers to either the specific set of satellites used in calculating a position, or all the satellites visible to a GPS receiver at one time, or the entire ensemble of GPS satellites comprising the Space Segment; Constellation refers to the network of orbiting global navigation satellites, with deference to their relative positions. The pattern created by the relative positioning of a navigation satellite network is designed to achieve a very high probability of global satellite coverage even in the event of satellite outages.

- **construction support**
offshore services related to the installation and construction of structures such as pipelines, drilling platforms and other oil and gas related infrastructure, usually involving the use of ROVs.

- **continuous kinematic surveying**
Successive baseline solutions generated at every epoch of an unbroken observation set. Typically used to track a vehicle or platform in motion.

- **contour interval**
The difference, in feet or meters, in elevation between two adjacent contour lines.

- **contour line**
A line on a map that connects points of equal elevation.

- **control point**
Also called a control station. A monumented point to which coordinates have been, or are being assigned by the use of surveying observations. The National Geodetic Survey maintains a nation-wide set of control points.

- **control segment**
A world-wide network of GPS monitoring and control stations that ensure the accuracy of satellite positions and their clocks;

A world-wide network of GPS monitoring and upload telemetry stations operated by, or on behalf of, the US Department of Defense. The tracking data is used by the Master Control Station at Colorado Springs to calculate the satellites' positions (or "broadcast ephemerides") and their clock biases. These are formatted into the Navigation Message which is uploaded on a daily (perhaps more frequently) basis by the Control Segment stations.

- **conus (contiguous united states)**

A term used to describe satellite coverage limited to the continental United States excluding Alaska and Hawaii.

- **conventional terrestrial system (cts)**

A standardized reference system, originating at the Earth's center of mass, designed to allow uniformity in geodetic measurements and computations. coordinated universal time (utc). Replaced Greenwich Mean Time (GMT) as the world standard for time in 1986. UTC uses atomic clock measurements to add or omit leap seconds each year to compensate for changes in the rotation of the earth.

- **coordinates**

A set of numbers that describes your location on or above the earth. Coordinates are typically based on latitude/longitude lines of reference or a global/regional grid projection (e.g., UTM, MGRS, Maidenhead); A set of numbers that represents a precise location anywhere on Earth. Usually stated as latitude and longitude.

- **correlation**

The extent to which one observation or computed value is influenced by the change in an other, or that both are influenced by a third (e.g. The error in a trigonometric height difference is correlated to the error in both the vertical angle and the measured distance. Similarly errors in computed GPS vectors have correlations between their individual components (X, Y, Z) and between common phase observations). The correlation coefficient is the proportion of the total variation in the dependent variable (y) which can be attributed to the relationship with the independent variable (x).

- **correlation-type channel**

A GPS receiver channel which uses a delay lock loop to maintain an alignment (correlation peak) between the replica of the GPS code generated in the receiver and the received code.

- **correlator**

The GPS receiver "software" or electronic means, implemented in some fashion (either analogue or digital) within a Tracking Channel, used to shift or compare the incoming signal with an internally generated signal. This operation is performed on the PRN codes, but may be used for more "exotic" mixed signals in the case of L2 measurements, where under the policy of Anti-Spoofing (AS) the L2 PRN code is not known. Correlator design may be influenced such that it is optimised for accuracy, mitigation of multipath, acquisition of signal under foliage, etc.

- **cors**
  Continuously Operating Reference Station Fixed GPS receiver site in continuous operation.

- **course**
  The direction from the beginning landmark of a course to its destination (measured in degrees, radians, or mils), or the direction from a route waypoint to the next waypoint in the route segment.

- **course deviation indicator (cdi)**
  A technique for displaying the amount and direction of cross track error (XTE): Shows your distance to the side of the desired course line.

- **course made good**
  The bearing from the "active from" waypoint to the current position, independent of the path taken to arrive at the current position.

- **course over ground**
  The current direction (in degrees) that you are actually traveling. (Selectable in degrees magnetic or true north)

- **course to steer**
  The heading you need to maintain in order to reach a destination.

- **course up orientation**
  Fixes the GPS receiver's map display so the direction of navigation is always "up."

- **covariance**
  The average value of the quantity x(r1) * x(r2); where x is a randomly varying function of the variable r, and r1 and r2 are two given values of r. Covariance describes the interdependence between variables and is typically expressed in a variance-covariance matrix where the diagonal elements are the variances of the corresponding variables, and those off the main diagonal are the covariance values.

- **cramming**
  A practice in which customers are billed for enhanced features such as voice mail, caller-ID and call-waiting that they have not ordered.

- **cross modulation**
  A form of signal distortion in which modulation from one or more RF carrier(s) is imposed on another carrier.

- **cross track error (xte)**
  The difference between a vessel's actual position and its desired position on a given heading. This is usually measured as a range error in nautical miles but may also be expressed graphically using symbols.

- **csu**
  Channel service unit. A digital interface device that connects end-user equipment to the local digital telephone loop. CSU is frequently coupled with DSU.

- **cts course to steer**
  The direction you should be traveling in order to return to the course while proceeding towards the destination. It is a "compromise" course bearing that projects from your current position to a point on the courseline midway between a point perpendicular to your position and the current leg destination waypoint.

- **cutoff angle**
  The minimum acceptable satellite elevation angle (above the horizon) to avoid blockage of line-of-sight, multipath errors or too high Tropospheric or Ionospheric Delay values. May be present in the receiver, or applied during data post-processing. For navigation receivers may be set as low as 5°, while for GPS Surveying typically a cutoff angle of 15° is used.

- **cycle slip**
  A discontinuity of an integer number of cycles in the measured (integrated) carrier phase resulting from a temporary loss-of-lock in the carrier tracking loop of a GPS receiver. This corrupts the carrier phase measurement, causing the unknown Ambiguity value to be different after the cycle slip compared with its value before the slip. It must be "repaired" (the unknown number of "missing" cycles determined and the carrier observation subsequent to the cycle slip all corrected by this amount) before the phase data is processed in double-differenced observables for GPS Surveying techniques; Instantaneous loss of lock (cycle count) on one or more carrier signals; A loss of continuity in the measured carrier beat phase which results from a temporary loss of lock on a GPS satellite; As a receiver measures the carrier frequency cycles of a global navigation satellite signal there may be periodic loss of signal due to atmospheric and other interferences. This is known as cycle slip.

- **dab (digital audio broadcasting)**
  Standard which enables the high-quality transmission of text, data and sound at a higher data rate.

- **dama**

Demand-Assigned Multiple Access - A highly efficient means of instantaneously assigning telephony channels in a transponder according to immediate traffic demands.

- **darc (data radio channel)**

A digital data transmission method using FM radio that is compatible with RDS and offers a higher bit rate (10 kbits/s)

- **darpa (defense advanced research projects agency)**

An agency of the USDoD (United States Department of Defense) responsible for the development of new technology for use by the military. DARPA, originally ARPA, was established in 1958. The agency is independent from other more conventional military R&D operations and focusses on short (two to four-year) projects run by small, purpose-built teams. ARPA was responsible for funding development of ARPANET (which grew into the Internet), as well as the Berkeley version of Unix and TCP/IP.

- **data**

Representation of facts, concepts, or instructions in a formalized manner suitable for communication, interpretation, or processing by humans or by automatic means.

- **data encryption standard (des)**

A cryptographic algorithm for the protection of unclassified computer data, issued as Federal Information Processing Standard Publication 46-1.

- **data links**

Data links are Differential Global Positioning System (DGPS) radio signals transmitted from a reference station or relay antenna at a known location to one or more rover receivers, to provide the positioning correction data that gives Thales Navigation GPS systems maximum accuracy, up to within a centimeter. Data links allow reference stations, relay transmitters and rover receivers to communicate calculations performed at each location to build a preponderance of data that is cross-checked at every point for verified precision. These data links are critical to collecting reliable GPS data. Thales Navigation has developed data link solutions for optimum system integration, and hardware and software compatibility with the highest market standards, to ensure superior performance and absolute reliability. A number of factors including range, update rate and error correction rate determine the quality and specific application of a data link. Thales Navigation has developed data link solutions that are readily adaptable to any application, overcoming the limitations to accuracy associated with vehicle tracking, DGPS, Local Area Differential GPS (LADGPS) and general data transfer. There are three types of data links deployed in Thales Navigation GPS systems for high-performance accuracy: 1) HF data link operates in the HF frequency between 1.6 MHz and 3.5 MHz. Primarily designed for LADGPS marine applications, HF data link is optimized for a meter-accurate update rate and easy government frequency allocation. Using a BCPSK modulation, advanced data compression, and high-performance GPS signal correction algorithms, HF data link offers superior accuracy for marine applications. With an operating range of 100 to 800 kilometer, HF data link is integrated into a number of Ashtech Marine Survey & Navigation GPS solutions. 2) UHF data link operates in the UHF frequency between 410 MHz and 470 MHz. Optimized for LADGPS or Real-Time Kinematic (RTK) marine and land survey applications, UHF data link is available in a number of configurations for maximum performance according to the requirements of specific Thales Navigation GPS systems. UHF data link can be configured with a DQPSK modulation, operating at 1200 baud. This configuration is compatible with earlier DGPS and Kinematic Applications in Real Time (KART) applications and uses NDS100MkII signal format. When configured with a GMSK modulation and Long Range Kinematic (LRK) signal format, UHF data link is optimized for LRK or KART kinematic applications. A third option uses a common signal format known as RTCM. UHF data link has a range of 80 kilometers and is integrated into a number of Ashtech Marine Survey & Navigation and Land Survey GPS solutions. 3) UHF bi-directional data link also operates in the UHF range and offers additional flexibility in configuring data processing functionality. Based on the Time Division Multiple Access (TDMA) principle, UHF bi-directional data link uses an MSK modulation optimized for relative positioning systems, used in marine seismic exploration and fleet management. UHF bi-directional data link is integrated into a number of Ashtech Marine Survey & Navigation and Fleet Management GPS solutions.

- **data message**

Also known as the Navigation Message. A 1500 bit message modulated on the L1 and L2 GPS signal, which contains the satellite's location (or ephemeris), clock (bias) correction parameters, constellation almanac information and satellite health; A 1500 bit message included in the GPS signal which reports the satellite's location, clock corrections, and health; A message included in the GPS signal that reports the satellite? location, clock corrections, and health. Included is approximate information about the other satellites in the system as well.

- **datalogger**

Also known as a Data Recorder. A handheld, lightweight data entry computer. It can be used to store additional data obtained by a GPS receiver, such as Attribute information on a Feature whose coordinates are captured for a GIS project.

- **datex (data exchange network)**

Agreement between European traffic centres to exchange traffic and travel information.

- **datum**

A math model which depicts a part of the surface of the earth. Latitude and longitude lines on a paper map are referenced to a specific map datum. The map datum selected on a GPS receiver needs to match the datum listed on the corresponding paper map in order for position readings to match; A Datum is a means by which coordinates determined by any means may be related to a well-defined Reference Frame. The Reference Frame may be visualised as a 3-D Cartesian coordinate system consisting, as a minimum, of information concerning the origin of the axes, and the directions of two principal axes fixed to the earth. The Reference Frame may be globally applicable, such as WGS84 or ITRF, in which case it is "geocentric" (having its origin at the earth's centre of mass), or be locally applicable as in the case of traditional national geodetic frames such as the Australian Geodetic Datum. In any case, the Datum may be considered synonymous to the Reference Frame, or be restricted to refer to the set of coordinates of geodetic stations or benchmarks which provide the physical realisation of the Reference Frame. A satellite-defined Datum such as WGS84 may, in addition, be realised by the time-varying coordinates of the satellites themselves (the Ephemerides). Finally, the Datum may be defined only in the horizontal sense or for the vertical component. An example of a Horizontal Datum is a Reference Ellipsoid (located and oriented in such a way as to be compatible to the Reference Frame to which it is attached), upon which coordinate information is expressed in terms of Latitude and Longitude. (WGS84 has a Reference Ellipsoid associated with it.) A Vertical Datum may be defined by a local realisation of Mean Sea Level, or as height above the Reference Ellipsoid; Map Datum's are the reference system used between the Lat/ Longs and the map being used as a reference. There are hundreds of them, and you have to tell your GPS unit which one to use. This site, Garmin and Delorme default to WGS84; A geodetic datum is a reference model of the earth, designed to fit all or part of the geoid (a model of the earth's surface based on gravity measurements). Datums provide the surface to which ground control measurements are referred, and are used as reference systems for navigation, surveying and mapping. All coordinates are referenced to a particular datum. World geodetic datums are typically defined by the size and shape of an ellipsoid, and the location of the center of the ellipsoid with respect to the center of the earth. The parameters that define a horizontal geodetic datum include the latitude and longitude of an initial point of reference (origin), and the radius and flattening of the ellipsoid selected for the computations. See also "Geodetic datum."; A reference system for vertical and horizontal positions. Different datums have different positions for the physical location of their origins, thus different datums will represent coordinates in different positions. Differences can be as much as a mile. All reliable maps that show coordinate systems provide Datum information.

- **datum, geodetic**

A set of constants specifying the coordinate system used for geodetic control. A complete geodetic datum provides, as a minimum, definition for orientation, scale and dimensions for the reference ellipsoid. The concept is generally expanded to include the published coordinates of control stations within the system.

- **db**

An analog unit of measure of signal strength, volume, or signal loss due to resistance as expressed in logarithmic form.

- **dbi**

The dB power relative to an isotropic source.

- **dbr**

Differential Beacon Receivers tune to the United States Coast Guard's high differential correction beacon stations for improved position accuracy. RTCM 104 input signal is supplied to a DGPS unit to correct for SA and Atmospheric errors. The operating frequency range is 285 kilohertz to 325 kilohertz

- **dbs**

  Direct broadcast satellite. Refers to service that uses satellites to broadcast multiple channels of television programming directly to home mounted small-dish antennas.

- **dbw**

  The ratio of the power to one Watt expressed in decibels.

- **dcii (digicipher ii)**

  An MPEG II digital standard developed by General Instruments Corporation which uses the MPEG II video encoding standards and the ATSC (Advanced Television Systems Committee) Dolby AC-3 digital audio standard. This standard has been adopted for standard and high definition television (SDTV & HDTV) transmission in North America, South Korea and Taiwan.

- **dde**

  Dynamic Data Exchange. A Windows protocol that allows communication between applications. When information is updated in one application, related information will be updated in other applications.

- **dead reckoning**

  A very simple method of using time and distance to navigate; The technique of determining position by computing distance traveled on a given course. Distance traveled is determined by multiplying speed by elapsed time; Process of determining a vehicle's approximate position by calculating from its last known GPS position a vector representing the movement since that point. Typically this is calculated using a combination of bearing and speed measurements from devices such as compasses and odometers.

- **decibel (db)**

  The standard unit used to express the ratio of two power levels. It is used in communications to express either a gain or loss in power between the input and output devices.

- **declination**

  The angular difference between True North and some other reference for north such as Grid North or Magnetic North; The offset angle of an antenna from the axis of its polar mount as measured in the meridian plane between the equatorial plane and the antenna main beam.

- **decoder**

  A television set-top device which enables the home subscriber to convert an electronically scrambled television picture into a viewable signal. This should not be confused with a digital coder/

decoder known as a CODEC which is used in conjunction with digital transmissions.

- **deemphasis**
  Reinstatement of a uniform baseband frequency response following demodulation.

- **default**
  The value or setting automatically chosen by the unit unless otherwise directed. Can be changed in Setup.

- **deflection of the vertical**
  The angle at a point on the surface of the earth between the vertical at that point (the line normal to the geoid) and the line through the point which is normal to the reference ellipse; The angle between the normal to the ellipsoid and the vertical (true plumbline). Since this angle has both a magnitude and a direction, it is usually resolved into two components: One in the meridian and the other perpendicular to it in the prime vertical.

- **de-gps: dbs-enhanced gps**
  Using a conventional GPS receiver in dense urban areas, a large portion of the sky is frequently obstructed by buildings, making it impossible to "see" a sufficient number of GPS satellites. The DE-GPS technique allows a GPS receiver to use the signal from a DBS satellite when it is in a bad spot. To do so, the DE-GPS receiver must obtain assistance information from a remote server. This assistance information enables the DE-GPS receiver to calculate the range information from the non-GPS satellites for the location fix. Background & Objective: GPS is essential for car navigation systems. However, when the car is in an area where most of the sky is blocked by buildings, it is difficult for the GPS receiver to acquire four satellites for a complete location fix. If additional satellites such as DBS can be used for geo-location, the probability of the receiver acquiring four satellites will be greatly increased. Technical Discussion: In dense cities, GPS receivers don't work well because tall buildings block GPS satellite signals. However, most urban areas are well covered by DBS satellites. DBS satellite signal can be used instead of GPS signal -- but assistance must be provided to receiver. Assistance is prepared by one reference station for all of Japan. Third Generation (3G) wireless service (e.g., FOMA) is used to send assistance from the reference station to the receiver. Additionally, the location fix can be more accurate than using only GPS because a DBS signal has a much wider signal bandwidth.

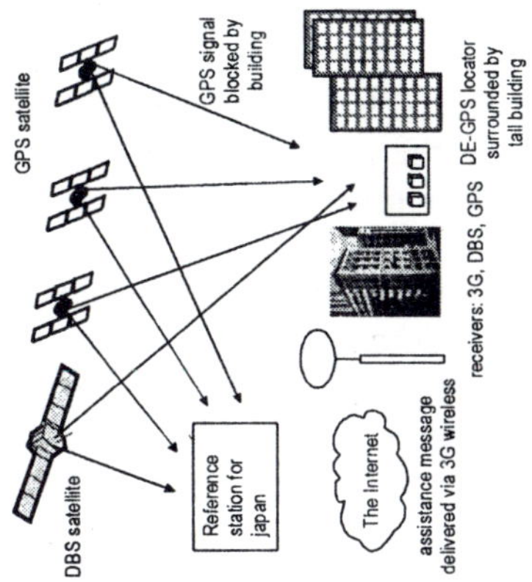

- **degree**

A unit of measurement; 360 degrees equal one complete circle, as around the earth on a line of latitude; measured from 0 to 90 degrees from the equator to each pole along a line of longitude.

- **degrees of freedom**

The number of observations minus the minimum number required to uniquely define the figure.

- **delay**

The time it takes for a signal to go from the sending station through the satellite to the receiving station. This transmission delay for a single hop satellite connection is very close on one-quarter of a second.

- **delay lock**

The technique whereby the received code (generated by the satellite clock) is compared with the internal code (generated by the receiver clock) and the latter is shifted in time until the two clocks match. Delay lock loops can be implemented in several ways; tau dither and early-minus-late gating.

- **delay-lock loop**

Another term for a code-tracking loop.

- **demand assigned multiple access (dama)**

Refers to contention access schemes that allow multiple communications users to share a discrete portion of the bandwidth.

- **demodulation**

The process for retrieving an information signal that has been modulated onto a carrier; Separating coded data from the carrier signal.

- **demodulator**

A satellite receiver circuit which extracts or "demodulates" the "wanted "signals from the received carrier.

- **depression**

An area in the surface of the earth that is lower in elevation than the surrounding area; a hole.

- **depression contour**

A contour line with hachures (small perpendicular tick marks) to signify a depression.

- **depth controlled gain (dcg)**
  A Garmin proprietary technology that automatically adjusts fishfinder sensitivity according to depth, not echo intensity. The result is a much more detailed and accurate picture of bottom structure.

- **descending node**
  The point on the ground track of the satellite orbit where the sub-satellite point(SSP) crosses the equator from the Northern Hemisphere into the Southern Hemisphere.

- **desired track (dtk)**
  The compass course between the "from" and "to" waypoints.

- **deviation**
  Errors from your course. Either built in or unintentional. One type of deviation is Magnetic, the difference between a true course and what a compass will tell you; The modulation level of an FM signal determined by the amount of frequency shift from the frequency of the main carrier.

- **dgn-microstation design file**
  A file format/extension.

- **dgps**
  GPS users can get improved accuracy with Differential GPS (DGPS), which corrects GPS signals to within an average of three to five meters (as compared to a typical 6-12 meters with the normal GPS signal). This system consists of a network of towers that receive GPS signals, and then transmit a corrected signal via beacon transmitters. In order to get the DGPS corrected signal, users must have a differential beacon receiver and beacon antenna in addition to their GPS; A Differential Global Positioning System (DGPS) is a system designed to improve the accuracy of Global Navigation Satellite Systems (GNSS) by measuring infinitesimal changes in variables to provide satellite positioning corrections. Two or more receivers observe the same set of satellites, taking similar measurements that produce similar errors when positioned closely together. A reference receiver, placed at a known location, calculates its theoretical position and compares it to the measurements provided by the navigation satellite signals. The difference between the two values reveals the measurement error. The reference receiver then transmits a corrected signal to any number of receivers at unknown positions within the area covered by the DGPS. Accuracy of global satellite positioning is thereby increased from 15 meters to within a few meters. This technique compensates for errors in the satellite navigation system itself but may not always correct errors caused by the local environment

when satellite navigation signals are reflected off of tall buildings or nearby mountains, creating multi-path signals. The accuracy of DGPS decreases with asynchronous measurement caused by spatial and temporal error decorrelation when the system receivers are set at greater distances apart.

- **dial around**

Long distance services that require consumers to dial a long-distance provider's access code (or "10-10" number) before dialing a long-distance number to bypass or "dial around" the consumer's chosen long-distance carrier in order to get a better rate.

- **differential (relative) positioning/differential correction**

Determination of relative coordinates of two or more receivers which are simultaneously tracking the same satellites. Dynamic differential positioning is a real-time calibration technique achieved by sending corrections to the roving user from one or more monitor stations. Static differential GPS involves determining baseline vectors between pairs of receivers.

- **differential gps (dgps)**

A technique to improve GPS accuracy that uses pseudo-range errors measured at a known Base Station location to improve the measurements made by other GPS receivers within the same general geographic area. It may be implemented in real-time through the provision of a communication link between the GPS receivers, transmitting the correction information in the industry-standard RTCM format, or various proprietary formats. May be implemented in single Base Station mode, in the so-called Local Area DGPS (LADGPS), or using a network of Base Stations, as in the Wide Area DGPS (WADGPS) implementation; A system devised initially by the U.S. Coast Guard that is now an international maritime standard to improve GPS accuracy levels to be within 10 meters. It employs a land based, fixed position, DGPS reference receiver to first calculate any basic GPS errors (and previously Selective Availability errors). It then transmits the necessary correction factors to mobile GPS receivers in the area. DGPS does require an added beacon receiver to communicate with the standard GPS unit; A method of improving GPS accuracy using measurements at a known location to improve the measurements made by other GPS receivers within the same general area.

• **differential gps positioning**

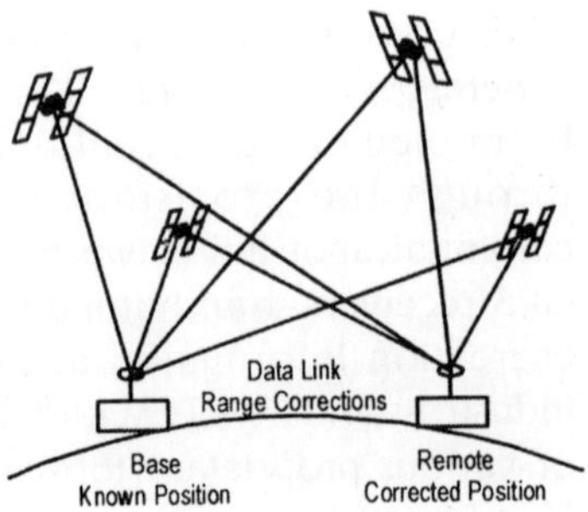

• **differential positioning**

Precise measurement of the relative positions of two receivers tracking the same GPS signals; Also known as Relative Positioning. Precise measurement of the relative positions of two receivers tracking the same GPS signals. May be considered synonymous with DGPS, or the term may be reserved for the more precise carrier phase-based baseline determination technique associated with GPS Surveying; GPS measurements can be differenced between receivers, satellites, and epochs. Although many combinations are possible, the present convention for differential processing of GPS phase measurements is to take differences between receivers (single difference), then between satellites (double difference), then between measurement epochs (triple difference).

(A) single-difference measurement between receivers is the instantaneous difference in phase of the signal from the same satellite, measured by two receivers simultaneously.

(B) double-difference measurement is obtained by differencing the single difference for one satellite with respect to the corresponding single difference for a chosen reference satellite.

(C) triple-difference measurement is the difference between a double difference at one epoch of time and the same double difference at the previous epoch of time.

• **digital**

The conversion of information into a binary format (1's and 0's or on's and off's), the smallest or simplest unit of information being a bit. These bits of information may be transmitted using radio frequency techniques such as terrestrial broadcasting, satellite and microwave transmission or over coaxial or fiber optic cable and copper wire. Digital information may be manipulated at extremely high speeds.

• **digital artifacts**

Errors in a digital transmission of data which result from digital compression applications which remove information from a data inventory in an amount sufficient to cause noticeable anomalies.

- **digital bit stream**
  A continuously changing flow of digital information which is being transported over some transmission medium such as cable or on a satellite microwave frequency.

- **digital compression**
  A process by which complex sets of algorithms are employed to rearrange, reassemble and eliminate digital information from a data inventory in such a way that the amount of data (bits) to be stored or transported is reduced without a resulting loss of apparent resolution.

- **digital data**
  Data represented by discrete values or conditions (i.e., "0" or "1"), as opposed to analog data.

- **digital raster graphic (drg)**
  The United States Geological Survey is releasing digital versions of all of the topographic maps. These files are called Digital Raster Graphics (DRG) maps. The images themselves are stored in a format called GeoTiff.

- **digital speech interpolation**
  DSI - A means of transmitting telephony. Two and One half to three times more efficiently based on the principle that people are talking only about 40% of the time.

- **digital tachograph**
  An electronic system fitted to commercial vehicles to record driver information and details about driving time and distance.

- **digital television (dtv)**
  A new technology for transmitting and receiving broadcast television signals. DTV provides clearer resolution and improved sound quality.

- **digital voice**
  Analog voice signals that have been converted into digital signals.

- **dilution of precision**
  The multiplicative factor that modifies ranging error. It is caused solely by the geometry between the user and his set of satellites. Known as DOP or GDOP; An indicator of satellite geometry for a unique constellation of satellites used to determine a position. Positions tagged with a higher DOP value generally constitute poorer measurement results than those tagged with lower DOP; A measure of the GPS receiver/ satellite geometry. A low DOP value indicates better relative geometry and higher corresponding accuracy. The DOP indicators are GDOP (geometric DOP), PDOP (position DOP), HDOP (horizontal DOP), VDOP (vertical DOP), and TDOP (time clock offset).

A value expressing the confidence factor in the accuracy of the position solution based on current satellite geometry. The lower the value the greater the confidence in the solution. DOP can be expressed in the following forms:

1. GDOP - All parameters are uncertain (latitude, longitude, height, clock offset).
2. HDOP - 2D parameters are uncertain (latitude, longitude).
3. HTDOP - 2D parameters and time are uncertain (latitude, longitude, time).
4. PDOP - 3D parameters are uncertain (latitude, longitude, height).
5. TDOP - Clock offset is uncertain.
6. VDOP - Height is uncertain.

A dimensionless number that accounts for the contribution of relative satellite geometry to errors in position determination. DOP has a multiplicative effect on the User Equivalent Range Error. Generally, the wider the spacing between the satellites being tracked by a GPS receiver, the smaller the position error.

The most common quantification of DOP is through the position dilution of precision (PDOP) parameter. PDOP is the number that, when multiplied by the root mean square (rms) UERE, gives the rms position error. Other DOPs include the geometric dilution of precision (GDOP), horizontal dilution of precision (HDOP), and vertical dilution of precision (VDOP); The multiplicative factor that modifies ranging error; Its sole cause is by the geometry between the user and his set of satellites. AKA DOP or GDOP.

- **direct broadcast satellite**

(DBS) Transmits TV signals directly to dishes at viewer's homes. Usually a high-powered satellite that requires only small dishes; A high-powered satellite that transmits or retransmits signals which are intended for direct reception by the public. The signal is transmitted to a small earth station or dish (usually the size of an 18-inch pizza pan) mounted on homes or other buildings.

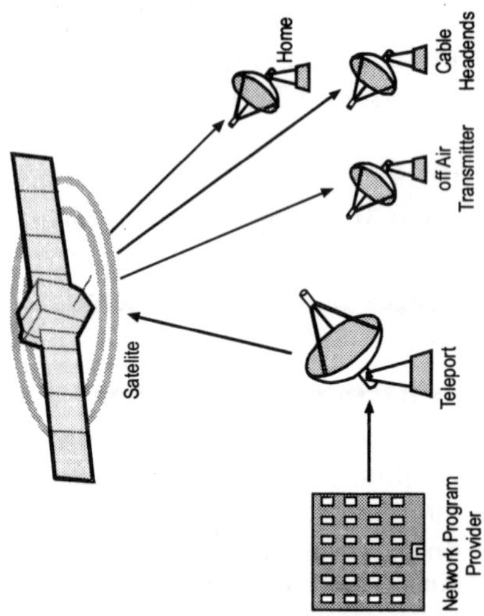

- **distance**
The length (in feet, meters, miles, etc.) between two waypoints or from your current position to a destination waypoint. This length can be measured in straight-line (rhumb line) or great-circle (over the earth) terms. GPS normally uses great circle calculations for distance and desired track.

- **distance root mean square (drms)**
A measurement used to describe the accuracy of a fix. It is twice the square root of the sum of the squares of all radial errors surrounding a true point divided by the total number of measurements.

- **distance to go (dtg)**
Digital readout (selectable in miles, nautical miles, or kilometers) displayed only when navigating to a waypoint. It simply indicates the remaining distance from your present position to the next waypoint.

- **dithering**
The introduction of digital noise into the system. "Clock dithering" is the process by which the U.S. Department of Defense (DoD) degrades the accuracy of the Standard Positioning Service (i.e. absolute positioning of a C/A-Code capable receiver). "Clock dithering" is the additional satellite clock "bias" induced by the DoD's "Selective Availability" policy that cannot be corrected for by the broadcast Navigation Message clock correction parameters.

- **dma-defense mapping agency**
Former name for the National Imagery and Mapping Agency (NIMA).

- **dnt/upt**
Count down-timer (DNT) - Starts at a user setting and counts down to zero. Useful for setting a time limit for travel. Count up-timer (UPT) - Starts at zero and counts upwards. Useful as a measure of elapsed time of travel.

- **dod**
United States Department of Defense. The government agency that controlled the development and deployment of the GPS system.

- **dop**
Dilution of Precision. Errors caused by bad geometry of the Satellites. The higher the number, the more "noise" in the position reading.

- **doppler aiding**
A signal processing strategy that uses a measured doppler shift to help the receiver smoothly track the GPS signal. Allows more precise velocity and position measurement.

- **doppler effect**
The shift in the frequency of a received radio signal due to the

relative motion of the transmitter and receiver.

- **doppler shift**
  The apparent change in the frequency of a signal caused by the relative motion of the transmitter and receiver; The change in frequency of a wave (such as RF emissions) caused by movement of the source relative to the observer; The apparent change in frequency of a received signal due to the rate of change of the range between the transmitter and receiver. See reconstructed carrier phase; A shift occurring in the electromagnetic spectrum, where an apparent change in signal frequency occurs as the transmitter and receiver move toward or away from one another.

- **doppler-aiding**
  A signal processing strategy that uses a measured Doppler Shift to help the receiver smoothly track the GPS signal. This allows for more precise velocity and position determination, especially when the receiver is moving at high speed and/or in an erratic fashion; The apparent change in the frequency of a signal caused by the relative motion of the transmitter and receiver.

- **doqq-digital orthophoto quarter-quadrangle**
  1:12,000 scale representations of a quarter of the area of a conventional USGS 7.5 minute topographical map.

- **double difference**
  Baseline computation algorithm using differences between carrier phase measurements between satellites and between receivers. This processing is the typical solution utilized for precise surveying applications. Computations are ambiguous in the unknown integer cycle term (see cycle ambiguity); A GPS observable formed by arithmetically differencing carrier phases simultaneously measured by a pair of receivers tracking the same pair of satellites. First, the phases obtained by each receiver from the first satellite are differenced. Second, the phases obtained by each receiver from the second satellite are differenced. And third, those differences are differenced. This procedure removes essentially all of the satellite and receiver clock errors. Although primarily used with carrier phases, the procedure can also be applied to pseudoranges; A data processing procedure by which the pseudo-range or carrier phase measurements made simultaneously by two GPS receivers are combined so that, for any measurement epoch, the observations from one receiver to two satellites are subtracted from each other (in a so-called

"between-satellite single-difference") to remove that receiver's clock error (or bias). (Similarly for the other receiver's observations to the same two satellites.) Then the two single-differences are subtracted so as to eliminate the satellite clock errors as well as to reduce significantly the effect of unmodelled atmospheric biases and orbit errors. (The order may be reversed, i.e., take "between-receiver single-differences" to each satellite in turn, and then difference between the single-differences.) The resulting set of Double-Differenced observables (for all independent combinations of two-satellite-two-receiver combinations) can be processed to solve for the baseline (linking the two receivers) components and, in the case of ambiguous carrier phase measurements, the integer ambiguity parameters. All high precision positioning techniques use some form of Double-Difference processing: pseudo-range, unambiguous carrier phase within a "bias-fixed" solution (i.e., after the double-differenced ambiguity values have been estimated and applied to the original carrier measurements), or ambiguous carrier phase data within a "bias-free" solution.

- **double-difference method**

  A method to determine that set of ambiguity values which minimized the variance of the solution for a receiver pair baseline vector.

- **downlink**

  A transmission path for the communication of signals and data from a communications satellite or other space vehicle to the earth.

- **dp (dynamic positioning)**

  an automatic pilot which controls a ship's engines and rudder, generally to ensure the vessel maintains station. Such systems require input from an accurate positioning system as a reference.

- **draw**

  A small, natural depression or valley in the side of a hill; a gully.

- **drg**

  Digital Raster Graphic. The United States Geological Survey is releasing digital versions of all of the topographic maps. These files are called Digital Raster Graphics (DRG) maps. The images themselves are stored in a format called GeoTiff.

- **drift**

  The rate of a vessel's departure from a given course as measured in knots.

- **dsl (digital subscriber line)**

  In ISDN (Integrated Services Digital Networks), equipment that provides full-duplex service on a single twisted metallic pair at a rate sufficient to support ISDN basic access and additional framing, timing recovery, and operational functions. The physical termination of the DSL at the network end is the line termination; the physical termination at the customer end is the network termination.

- **ds-level**

  Digital Signal Level:

  1. DS-0: 64 Kb/s.
  2. DS-0A: A DS-0 signal containing data from a single subrate station. The bytes are repeated as necessary to match the sender's speed to the speed of the DS-0 line.
  3. DS-0B: A DS-0 signal containing data from several subrate stations. The octets contain a subrate synchronizing bit (bit# 1) and six data bits (bits 2 through 7). The eighth bit is set to 1 to ensure meeting the 1s density requirement.
  4. DS-1: 1.544 Mb/s.
  5. DS-1C: 3.152 Mb/s.
  6. DS-2: 6.312 Mb/s.
  7. DS-3: 44.736 Mb/s.
  8. DS-4: 274.176 Mb/s.
  9. DS-4A: 139.264 Mb/s.

- **dsrc (dedicated short range communication)**

  Standard enabling data transmission at a frequency of 5.8 GHz.

- **dst distance**

  Distance from position to destination.

- **dsu**

  Data service unit. A device used in digital transmission that adapts the physical interface on a DTE device to a transmission facility such as T1 or E1. The DSU is also responsible for such functions as signal timing. DSU is frequently coupled with a CSU as CSU/DSU.

- **dtam**

  Digital world Tectonic Activity Map.

- **dtk**

  Desired Track. The course between to and from.

- **dtm-digital terrain model**

  A digital model of the terrain surface-similar to a DEM but often enhanced by planimetric information.

- **dual spin**

  Spacecraft design whereby the main body of the satellite is spun to provide altitude stabilization, and the antenna assembly is despun by means of a motor and bearing system in order to continually direct the antenna earthward. This dual-spin configuration thus serves to create a spin stabilized satellite.

- **dual-frequency**

  Refers to the instrumentation that can make measurements on both

L-Band frequencies, or to the measurements themselves (e.g., L1 and L2 pseudo-range or carrier phase measurements). Dual-frequency measurements are useful for high precision (pseudo-range-based) navigation because the Ionospheric Delay bias can be determined, and the data corrected for it. In the case of Double-Differenced carrier phase, dual-frequency observations can account for the residual ionospheric bias (for case of long baselines), or aid Ambiguity Resolution for "rapid static" or "kinematic" baseline determination. All "top-of-the-line" GPS receivers are of the dual-frequency variety, and are comparatively expensive because of the special signal processing techniques that must be implemented to make measurements on the L2 carrier under the policy of Anti-Spoofing.

- **duplex transmission**

Capability for simultaneous data transmission between a sending station and a receiving station. Most modems have a switch that lets you choose between full-duplex and half-duplex modes. The choice depends on which communications program you are running. In full-duplex mode, data you transmit does not appear on your screen until it has been received and sent back by the other party.

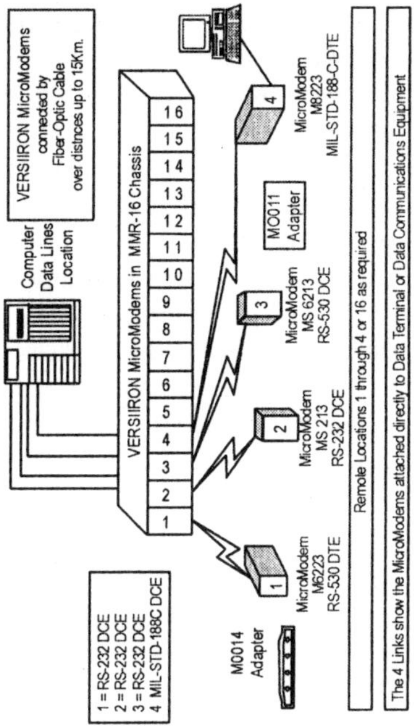

- **dvb**

A suite of digital video standards formulated by the European Launch Group (ELG). The standard is based on MPEG II video. MPEG II/DVB uses the Musicam audio standard.

- **dvd-rom**

A single DVD-ROM disc contains the mapping data of the entire North American sub-continent. Systems with DVD capability don't require you to change discs depending on your location.

- **dxf-drawing exchange format**

Used for transfering CAD (Computer Aided Drafting) files.

- **dynamic positioning**
Determination of a timed series of sets of coordinates for a moving receiver, each set of coordinates being determined from a single data sample, and usually computed in real time; Dynamic positioning or kinematic positioning refers to applications in which the position of a moving object is determined; The process of collecting GPS data while the GPS antenna is in motion. Often associated with Line or Area Features.

- **e1**
Wide-area digital transmission facility used predominantly in Europe that carries data at a rate of 2.048 Mbit/s.

- **e3**
Wide-area digital transmission facility used predominantly in Europe that carries data at a rate of 34.368 Mbit/s.

- **earth centered, earth fixed (ecef)**
A Cartesian coordinate system beginning at the Earth's center of mass. The Z-axis is aligned with the Earth's mean spin axis. The X-axis is aligned with the zero meridian. The Y-axis is 90 degrees west of the X-axis, forming a right-handed coordinate system.

- **earth station**
The term used to describe the combination or antenna, low-noise amplifier (LNA), down-converter, and receiver electronics. used to receive a signal transmitted by a satellite. Earth Station antennas vary in size from the.2 foot to 12 foot (65 centimeters to 3.7 meters) diameter size used for TV reception to as large as 100 feet (30 meters) in diameter sometimes used for international communications. The typical antenna used for INTELSAT communication is today 13 to 18 meters or 40 to 60 feet.

- **eccentricity**
The orbital parameter used to describe the geometric shape of an elliptical orbit; eccentricity values vary from (e = 0) to (e = 1), where (0) describes a circle and (1) describes a straight line

- **ecdis-electronic charting display information system**
A system with sufficient precision that commercial navigation can be accomplished without the aid of paper charts.

- **ecef coordinates**
Cartesian coordinates, Earth Centered Earth Fixed Rectangular coordinate system where the positive X axis lies on the equatorial plane passing through the prime meridian (Greenwich), the positive Y axis lies on the equatorial plane at 90 east, and positive Z passes north through the mean rotational axis of the earth.

- **echo canceller**
An electronic circuit which attenuates or eliminates the echo effect on satellite telephony links. Echo cancellers are largely replacing obsolete echo suppressors.

- **echo effect**
A time-delayed electronic reflection of a speaker's voice. This is largely eliminated by modern digital echo cancellers.

- **echo suppression**
Refers to the cancellation of any reflected signal in a communications channel.

- **edge of coverage**
Limit of a satellite's defined service area. In many cases, the EOC is defined as being 3 dB down from the signal level at beam center. However, reception may still be possible beyond the 3dB point.

- **edi (electronic data interchange)**
A standard form of electronic communication used mainly for electronic commerce and document interchange.

- **efc (electronic fee collection)**
A system that enables automatic debiting/payment for a transport service without any action from the user at the moment of use - example: automatic toll collection.

- **egnos (european geostationary navigation overlay service)**
A European system of satellite navigation. It will augment the two satellite navigation systems now operating, the U.S. GPS and GNSS systems. It consists of three geostationary satellites and a network of ground stations. EGNOS will achieve its aim by transmitting a signal containing information on the reliability and accuracy of the positioning signals sent out by GPS and GNSS.

- **eirp**
Effective Isotropic Radiated Power - This term describes the strength of the signal leaving the satellite antenna or the transmitting earth station antenna, and is used in determining the C/N and S/N. The transmit power value in units of dBW is expressed by the product of the transponder output power and the gain of the satellite transmit antenna.

- **electromagnetic spectrum**
The continuous distribution of energy in the form of electromagnetic waves, which are arranged in order of their frequencies or wavelengths; The entire range of frequencies of electromagnetic radiation from zero to infinity which measure electromagnetic energy. The Radio spectrum (less than 300 GHz bandwidth) is divided into subsets

of various bandwidths which are identified by their frequency characteristics. For example, the Ku-band subset includes those frequencies of the radio frequency (RF) spectrum between 10.7 GHz and 17.8 GHz while the C-band subset includes those frequencies between 3600 MHz and 7075 MHz.

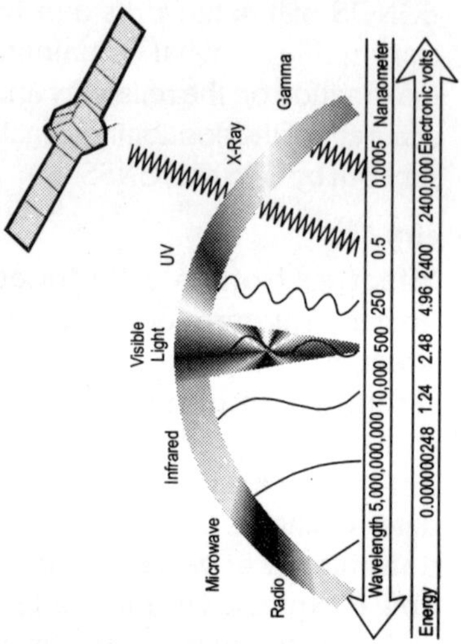

- **electronic access**

  The capability to access information via on-line access (dedicated or dial-up), E-Mail, and FAX.

- **electronic data interchange (edi)**

  The exchange of routine business transactions in a computer-processable format, covering such traditional applications as inquiries, planning, purchasing, acknowledgments, pricing, order status, scheduling, test results, shipping and receiving, invoices, payments, and financial reporting.

- **elevation**

  Height above mean sea level or vertical distance above the reference geoid; The vertical distance of a point above or below a reference surface, such as sea level; The upward tilt to a satellite antenna measured in degrees required to aim the antenna at the communications satellite. When. aimed at the horizon, the elevation angle is zero. If it were tilted to a point directly overhead, the satellite antenna would have an elevation of 90 degrees.

- **elevation mask**

  Elevation mask is an adjustable feature of GPS receivers that specifies a satellite must be a certain number of degrees above the horizon before its signals are used for positioning. Satellites at low elevation angles (five degrees or less) have lower signal strengths and are more prone to loss of lock thus causing noisy solutions.

- **elevation mask angle**

  The elevation angle below which satellites are ignored in the position calculation. Typically set to around 5-10 degrees to avoid interference problems from obstructions and atmospheric errors.

- **ellipsoid**

A geometric surface, all of whose plane sections are either ellipses or circles; Spheroid: The mathematical function used to describe the shape of the earth for geodetic computations. The figure is formed by rotating an ellipse about its minor (shorter) axis and is typically described by dimensions for the semimajor axis (a) together with the semiminor axis (b) or flattening (f) = (a-b)/a; A smooth, mathematical surface closely approximating the size and shape of the earth. In geodesy, unless otherwise specified, a mathematical figure formed by revolving an ellipse about its minor axis. It is often used interchangeably with spheroid. Two quantities define and ellipsoid; these are usually given as the length of the semimajor axis, a, and the flattening, f = (a-b)/a, where b is the length of the semiminor axis. Prolate and triaxial ellipsoids are invariably described as such.

- **ellipsoid height**

The measure of vertical distance above the ellipsoid. Not the same as elevation above sea level. GPS receivers output position-fix height in the WGS-84 datum.

- **ellipsoid of revolution**

Often referred to simply as ellipsoid, ellipsoid of revolution is a mathematical representation of the Earth as an ellipse that is rotated about its minor axis. An ellipsoid is an equipotential surface of a rotating, homogeneous body. Various ellipsoid models have been determined to approximate the geoid in local areas and in a global sense. GPS uses the WGS84 earth model which is based on the GRS80 ellipsoid.

- **elliptical orbit**

Those orbits in which the satellite path describes an ellipse with the Earth at one focus.

- **e-mail**

Also called electronic mail, refers to messages sent over the Internet. E-mail can be sent and received via newer types of wireless phones, but you generally need to have a specific e-mail account.

- **en banc**

An informal meeting held by the Commission to hear presentations on specific topics by diverse parties. The Commissioners, or other officials, question presenters and use their comments in considering FCC rules and policies on the subject matter under consideration.

- **en route**

Navigation between the point of departure and point of arrival.

- **encapsulator**

An IP gateway at a satellite uplink or cable/wireless headend which is a router taking traffic from the terrestrial network and routing packets to the satellite WAN (Wide Area Network). The encapsulator offers streaming capabilities that encapsulate content into datagrams and then route it to subscriber subnets.

- **encoder**
  A device used to electronically alter a signal so that it can only be viewed on a receiver equipped with a special decoder.
- **encrypt**
  To convert plain text into an unintelligible form by means of a cryptosystem.
- **encryption**
  The action of disguising information so that it can be recovered relatively easily by persons who have the key, but is highly resistant to recovery by persons who do not have the key. Encryption is accomplished by scrambling the bits, characters, works or phrases in the original message.
- **enhanced service providers**
  A for-profit business that offers to transmit voice and data messages and simultaneously adds value to the messages it transmits. Examples include telephone answering services, alarm/security companies and transaction processing companies.
- **eol**
  End of Life of a satellite.
- **epe**
  Estimated Position Error. How much the unit thinks it is off target.
- **epe-estimated position error**
  The distance the GPS receiver may be off target.
- **ephemeris**
  The predictions of current satellite position that are transmitted to the user in the data message; The description of the satellite orbits and clock correction parameters variable over time used for positioning and baseline computations. The ephemerise may be broadcast (projected ahead into time and subject to selective availability) or precise (post-fitted); The predictions of current satellite position that are transmitted to the user in the data message; It is a Set of satellite orbital parameters used by a GPS receiver to calculate precise GPS satellite positions and velocities. The ephemeris is used in the determination of the navigation solution and is updated periodically by the satellite to maintain the accuracy of GPS receivers. Available as a broadcast ephemeris or as a post-processed precise ephemeris; A list of (accurate) positions or locations of a celestial object as a function of time. Available as

"broadcast ephemeris" or as preprocessed "precise ephemeris."; Ephemeris is a set of parameters used by a global navigation satellite receiver to predict the location of a satellite and its clock behavior. Each satellite contains and transmits ephemeris data about its own orbit and clock. Ephemeris data is more accurate than the almanac data but is applicable over a short time frame from four to six hours. Ephemeris data is transmitted by the satellite every 30 seconds.

- **ephemeris data**
  Current satellite position and timing information transmitted as part of the satellite data message. A set of ephemeris data is valid for several hours.

- **ephemeris errors**
  Errors (or "biases") which are present in the (Broadcast or Precise) Ephemeris data. Broadcast Ephemeris errors are typically at the few metre level, while Precise Ephemeris errors are at the decimetre-level. Ephemeris errors are largely mitigated by differential correction (in DGPS Positioning) or in double-differenced observables (formed from carrier phase measurements) when the receivers are not up to a few tens of kilometres apart. In very high precision applications and/or where the baseline lengths are hundreds or thousands of kilometres, residual Ephemeris Errors may limit the accuracy of the baseline solution.

- **epoch**
  A specific instant in time. GPS carrier phase measurements are made at a given frequency (e.g. every 30 seconds) or epoch rate; It is measurement interval or frequency at which data is measured. For example a GPS receiver reporting a position once every two seconds as an epoch of two seconds; Measurement interval or data frequency, as in making observations every 15 seconds. Loading data using 30-second epochs means loading every other measurement.

- **epoch date**
  The date, usually expressed in decimal years, for which published coordinates and data are valid.

- **equator**
  Zero degrees Latitude. A line around the center of earth 24,901.55 miles (40,075.16 kilometers) long; The Latitude reference point for the Lat/Long grid system. Other positions of latitude are referenced as degrees of North Latitude if they are north of the equator and degrees of South Latitude if they are south of the equator. Also see Latitude.

- **equatorial orbit**

An orbit with a plane parallel to the earth's equator.

- **equivalent noise temperature**

The temperature, usually expressed in Kelvins, of a hypothetical matched resistance at the input of an assumed noiseless device, such as a noiseless amplifier, that would account for the measured output noise.

- **error budget**

A set of individual error sources with statements of the percentage of the total system error contributed by each source.

- **error ellipse**

A statistical measure of the positional error at a given point computed from the propagation of all errors contributing to the position and expressed by its semi-major and semi-minor axis (vectors of greatest and least magnitude) and the covariance (rotation angle from the reference coordinate system). Two dimensional errors are typically propagated at one-standard deviation (39.4% probability that the position lies on or within the ellipse) or 2.1447 times the standard deviation (95% confidence) level; A statistical measure of the positional error at a given point computed from the propagation of all errors contributing to the position.

- **esc**

Engineering Service Circuit - The 300-3,400 Hertz voice plus teletype (S+DX) channel used for earth station-to-earth station and earth station-to-operations center communications for the purpose of system maintenance, coordination and general system information dissemination. In analog (FDM/FM) systems there are two S+DX channels available for this purpose in the 4,000-12,000 Hertz portion of the baseband. In digital systems there are one or two channels available which are usually convened to a 32 or 64 Kbps digital signal and combined with the earth station traffic digital bit stream. Modern ESC equipment interfaces with any mix of analog and digital satellite carriers, as well as backhaul terrestrial links to the local switching center.

- **esp (electronic stability program)**

Safety function linked to brake sensors and traction control which enables cars to correct their trajectory on slippery roads and curves.

- **estimated position error (epe)**

A measurement of horizontal position error in feet or meters based upon a variety of factors including DOP and satellite signal quality; An estimation of the accuracy of your position that tells you how far off target your unit might be. This calculation is based on the geometry and position of tracked satellites, time clock offset, satellite signal quality, and more.

- **estimated time enroute (ete)**
The time it will take to reach your destination (in hours/minutes or minutes/seconds) based upon your present position, speed, and course; An estimation of the time required to travel to a pre-selected waypoint or the next waypoint in a route. This calculation is based on the speed and direction of travel towards the selected waypoint. The time left to your destination at your present speed. Typically used for navigation applications.

- **estimated time of arrival (eta)**
The time of day of your arrival at your destination.

- **ethernet**
A standard protocol (IEEE 802.3) for a 10 Mb/s baseband LAN (local area network) bus using carrier-sense multiple access with collision detection (CSMA/CD) as the access method, implemented at the Physical Layer in the ISO Open Systems Interconnection-Reference Model, establishing the physical characteristics of the CSMA/CD network.

- **ethernet cable**
Cable transmission medium for Ethernet classified as XbaseY, where X is the data rate in Mb/s, "base" means baseband (as opposed to radio frequency) and Y is the category of cabling. The original cable was 10base5. Common XbaseY Ethernet Cable Classifications include:

1. **10baseT:** A variant of Ethernet which allows stations to be attached via twisted pair cable.
2. **10base2:** The variant of Ethernet that uses thin coaxial cable (RG-58 or similar) limited to a maximum single cable length of 200 meters; (sometimes referred to as "cheapernet".)
3. **10base5:** The original "full spec" variant of Ethernet cable, using a stiff, large diameter coaxial with an impedance 50 ohms and with multiple shielding. Maximum length is limited to 500 meters (Sometimes referred as "thicknet" or "thick Ethernet").
4. **100BaseT:** Any several "Fast Ethernet" 100 Mb/s, CSMA/CD (Carrier Sense Access/ Collision Detect) standards for twisted pair cables including: 100BaseTx (100 over two-pair Category 5 or better cable) 100 BaseT4(100 mb/s over four-pair Category 3 or better cable and 100BaseT2

(100 Mb/s over two-pair Category 3 or better cable). All are standards under IEEE 802.3.

5. **100Base FX**: "Fast Ethernet" over optical fibre. 6. 100Base VG: A 100 Mb/s Ethernet standard specified to run over four pairs of Category 3 UTP(Unshielded Twisted Pair) wires (known as Voice Grade or VG). It is also called 100VG-AnyLAN because it was defined to carry both Ethernet and token ring frame types.

- **faraday rotation**

Rotation of the plane of polarization of an electromagnetic wave when traveling through a magnetic field. In space communications, this effect occurs when signals traverse the ionosphere.

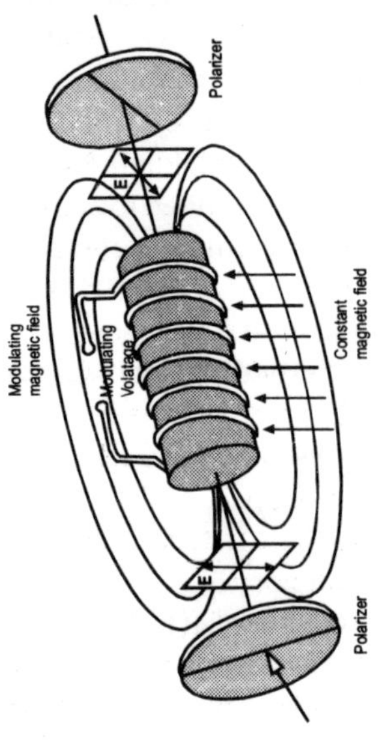

- **fast ambiguity resolution**

Rapid static, Fast static GPS surveying technique utilizing multiple observables (dual-frequency carrier phase, C/A and P codes) to resolve integer ambiguities with shortened observation periods. The method may also be used for observations with the receiver in motion known as on-the-fly ambiguity resolution.

- **fast switching channel**

A single channel which rapidly samples a number of satellite ranges. "Fast" means that the switching time is sufficiently fast (2 to 5 milliseconds) to recover the data message.

- **fastrack™**

A selectable on/off feature, FasTrack™ displays flasher-style sonar returns in a narrow vertical column at the far right of the sonar screen. Especially useful for ice fishing or fishing at anchor, FasTrack™ is a standard feature on all Lowrance sonar units for 2001.

- **fcc (federal communications commission)**

This is the national regulatory body for interstate tele-communications in the United States. The commission consists of five members all nominated to a specific term by the President of the United States and confirmed by the Senate. The current chairperson is William Kennard. The authority for the commission is contained within the Communications Act of 1934 as amended.

- **fcd (floating car data)**

Floating car data are gathered from vehicles equipped with appropriate sensors (such as whether windscreen wipers are activated or the vehicle's location and speed) and then passed to traffic management centres that can evaluate this information for general traffic trends.

- **fdma**

Frequency division multiple access. Refers to the use of multiple carriers within the same transponder where each uplink has been assigned frequency slot and bandwidth. This is usually employed in conjunction with Frequency Modulation.

- **feature**

The object which is being mapped for use in a GIS system. Features may be points, lines or areas; The process of collecting GPS and GIS information simultaneously.

- **fec (forward error correction)**

A technique which employs special codes that allow the receiver to detect and correct a limited number of errors without referring to the transmitter.

- **federal radionavigation plan (frp)**

Congressionally managed, joint Department of Defense (DoD) and Department of Transportation (DOT) effort to reduce the proliferation and overlap of federally funded radionavigation systems. The FRP is designed to delineate policies and plans for U.S. government-provided radionavigation services.

- **feed**

This term has at least two key meanings within the field of satellite communications. It is used to describe the transmission of video programming from a distribution center. It is also used to describe the feed system of an antenna. The feed system may consist of a subreflector plus a feedhorn or a feedhorn only.

- **fgdc**

The Federal Geographic Data Committee (FGDC) is an interagency committee, organized in 1990 under OMB Circular A-16 that promotes the coordinated use, sharing, and dissemination of geospatial data on a national basis. The FGDC is composed of representatives from nineteen Cabinet level and independent federal agencies. The Steering Committee sets high-level strategic direction for the FGDC as a whole. The Coordination Group advises on the day-to day business of the

FGDC. Staff support for FGDC committees is provided by the FGDC Secretariat staff; The Federal Geographic Data Committee (FGDC) is an interagency committee, organized in 1990 under OMB Circular A-16 that promotes the coordinated use, sharing, and dissemination of geospatial data on a national basis.

- **fiber optics**

  An informational transfer technology using transparent fibers of glass or plastic to transmit light from one point to another.

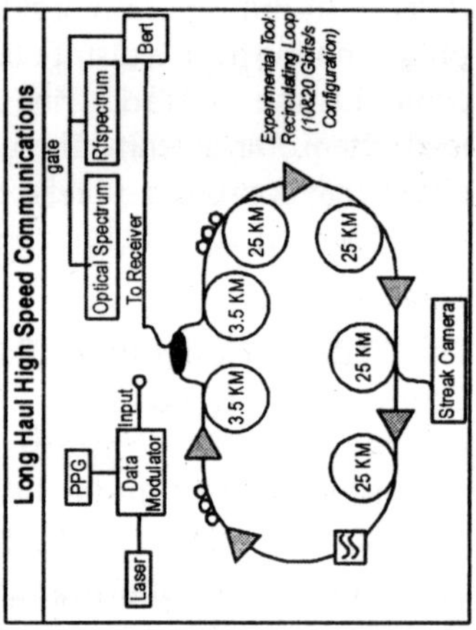

- **film supertwist**

  Many Lowrance fish finder and GPS/mapping units use this technology for liquid crystal displays (LCD). Film SuperTwist provides excellent contrast and readability...in direct sunlight or with the backlight for fishing at night or in low light...and from virtually any viewing angle. Lowrance Film SuperTwist screens are also easy to see with polarized sunglasses.

- **firmware**

  Firmware is the electronic heart of a receiver, where coded instructions relating to receiver function, and sometimes data processing algorithms, are embedded as integral portions of the internal circuitry.

- **fix**

  A single position with latitude, longitude (or grid position), altitude, time, and date; A position that is determined by the navigation unit (consisting of latitude, longitude (or grid position), altitude, time, and date); Defining position, where two lines such as latitude and longitude cross, as determined by one or more navigation aids or techniques.

- **fjord-(also fiord)**

  A long, narrow inlet of the sea between steep slopes.

- **flashgraf™**

  FlashGraf™ combines a scrolling sonar target graph with a digital LCD flasher with GRAYLINE® or COLORLINE™. This allows you to view two of the most popular ways of finding fish all in one screen. FlashGraf™ also employs the "Resize Window" feature, which allows you to make the sonar graph or flasher "window" wider or narrower in the split-screen display according to your preference. Also, the "Overlay Data" feature is included to allow temp, speed, lat/long, and more information to be displayed in the flasher "window".

- **flattening**

  A parameter used to define the shape of an ellipsoid. The WGS-84 ellipsoid has a flattening of 1/298.257223563.

- **fli-map**

  a system that, with the help of a laser fan beam in a helicopter, generates accurate relief maps.

- **floatducer™**

  This is a unique, floating transducer you can use when fishing off a bank, shore, bridge or dock. Shaped like an odd-looking UFO, just toss the FloatDucer™ onto the water with its 50' cable connected to a fixed-mount or portable Lowrance sonar. FloatDucer™ provides an effective wide-angle underwater coverage area of up to 60 degrees with ASP™ on.

- **fluxgate compass**

  An electronic device used to measure magnetic direction.

- **fm**

  Frequency Modulation. A method of encoding information about a carrier signal by altering the frequency while amplitude remains constant; Modulation in which the instantaneous frequency of a sine wave carrier is caused to depart from the center frequency by a amount proportional to the instantaneous value of the modulating signal. In FM, the carrier frequency is called the center frequency. In optical communications, even if the electrical baseband signal is used to frequency-modulate an electrical carrier, it is still the intensity of the light wave that is varied (modulated) by the electrical FM carrier. The light wave is varied in intensity at an instantaneous rate corresponding to the instantaneous frequency of the electrical carrier.

- **focal length**

  Distance from the center feed to the center of the dish.

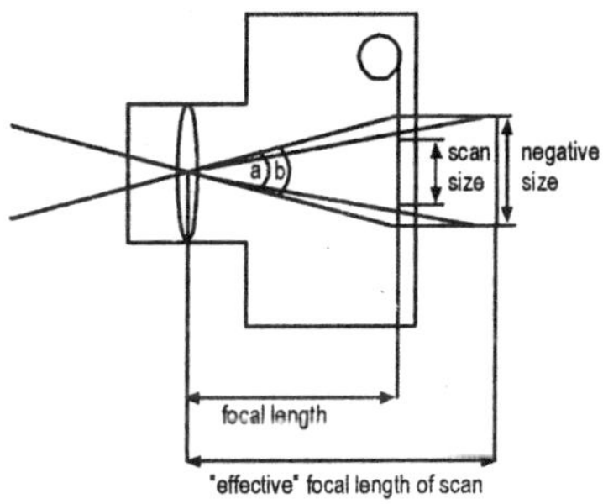

- **focal point**

  The area toward which the primary reflector directs and concentrates the signal received.

- **footprint**

  A map of the signal strength showing the EIRP contours of

equal signal strengths as they cover the earth's surface. Different satellite transponders on the same satellite will often have different footprints of the signal strength. The accuracy of EIRP footprints or contour data can improve with the operational age of the satellite. The actual EIRP levels of the satellite, however, tends to decrease slowly as the spacecraft ages.

- **ford**

A shallow place in a body of water, such as a river, where one can cross on foot, by riding an animal, or in a vehicle.

- **form factor**

A term used to describe the way a device is constructed, such as its shape. For example, GPS receivers come in many form factors, such as handheld, PDA, automotive, USB, etc.

- **forward error correction (fec)**

Adds unique codes to the digital signal at the source so errors can be detected and corrected at the receiver.

- **frequency**

The number of repetitions per unit time of a complete waveform, as of a radio wave (see L1 and L2 frequencies in this glossary); The number of waves passing a specific point within a unit period of time, expressed in Hertz (cycles per second); The number of times that an alternating current goes through its complete cycle in one second of time. One cycle per second is also referred to as one hertz; 1000 cycles per second, one kilohertz; 1,000,000 cycles per second, one megahertz: and 1,000,000,000 cycles per second, one gigahertz.

- **frequency band**

A range of frequencies in a particular region of the electromagnetic spectrum.

- **frequency coordination**

A process to eliminate frequency interference between different satellite systems or between terrestrial microwave systems and satellites. In the U.S. this activity relies upon a computerized service utilizing an extensive database to analyze potential microwave interference problems that arise between organizations using the same microwave band. As the same C-band frequency spectrum is used by telephone networks and CATV companies when they are contemplating the installation of an earth station, they will often obtain a frequency coordination study to determine if any problems will exist.

- **frequency division multiple access(fdma)**

Multiple access being achieved through frequency band

discrimination; either in the ground receiver, in the case of a wide band frequency translating repeater, or in the satellite, in the case of a frequency channelized satellite.

- **frequency modulation**
  A method of encoding information about a carrier signal by altering the frequency while amplitude remains constant.

- **frequency spectrum**
  The distribution of signal amplitudes as a function of frequency.

- **fss (fixed satellite service)**
  The segment of Ku-band satellite service established by the FCC to be provided from medium power satellites. These satellites are separated in orbit by at least 2 degrees. RF signals are transmitted to FSS satellites in the 14 GHz to 14.5 GHz range and received from the satellite in the 11.7 GHz to 12.2 GHz range.

- **fubar**
  Fouled Up Beyond All Repair.

- **full operational capability (foc)**
  For NAVSTAR, FOC occurred in 1994, when 24 operational GPS satellites (Block II/IIA) began functioning in their assigned orbits and were available for navigation use.

- **full-duplex operation**
  A mode of operation in which simultaneous communication in both directions may occur between two terminals. Contrast with half duplex or simplex operation in which communications occur in only one direction at a time.

- **fullwave**
  Term used to differentiate between measurements made with signal-squared (codeless) and code-tracking receivers. Specifically, a receiver tracking L2 P-code can make measurements using the whole L2 wavelength (23 cm): the full wave.

- **fundamental frequency**
  The fundamental frequency used in GPS is 10.23 MHz. The carrier frequencies L1 and L2 are integer multiples of this fundamental frequency. L1 = 154F = 1575.42 MHz L2 = 120F = 1227.60 MHz

- **g.p.s.-global positioning system**
  A generic term for satellite based positioning system.

- **g/t**
  A figure of merit of an antenna and low noise amplifier combination expressed in dB. "G" is the net gain of the system and "T" is the noise temperature of the system. The higher the number, the better the system.

- **g/t (gain to noise temperature)**
  In the characterization of antenna performance, a figure of merit, where G is the antenna gain in dB (decibels) at the receive

frequency, and T is the equivalent noise temperature of the receiving system in Kelvin.

- **gain**

The ratio of output current, voltage or power to input current, voltage or power, respectively. Gain is usually expressed in dB. If the ratio is less than unity, the gain expressed in dB, will be negative, in which case there is a loss between input and output.

- **gans**

The Air Force's Global Access, Navigation, and Safety (GANS) program.

- **gans-global access, navigation, and safety**

A United States Air Force program that is a potential vehicle for collaboration. GANS is an umbrella avionics program that integrates GPS, navigation and safety equipment, Navigation Warfare (NAVWAR), avionics modernization, military ground-based infrastructure, Global Air Traffic Management (GATM), and the Joint Precision Approach and Landing System (JPALS).

- **gateway**

In a communication network, one of the network nodes equipped for interfacing with a network using different protocols. Note 1: A gateway may contain devices such as protocol translators, impedance matching devices, rate converters, fault isolation, or signal translators as necessary to provide system interoperability.

- **gats (global automotive telematics standard)**

Open standard to enable a full range of traffic information and other telematics services.

- **gdop**

Geometric Dilution of Precision, see DOP.

1. GDOP Components
2. PDOP = Position Dilution of Precision (3-D), sometimes the Spherical DOP.
3. HDOP = Horizontal Dilution of Precision (Latitude, Longitude).
4. VDOP = Vertical Dilution of Precision (Height).
5. TDOP = Time Dilution of Precision (Time).

Geometric Dilution of Precision. The relationship between errors in user position and time and in satellite range.

GDOP2 = PDOP2 + TDOP2

See PDOP.

- **geo (geosynchronous earth orbit)**

This is the orbital altitude of 35,580 km (22,237 miles) above the earth's surface where a satellite's velocity matches with the rotation of the earth. A satellite which is in a GEO position above the earth's equator (geostationary) will appear from the earth to be occupying a stationary position. The geosynchronous earth orbit is also referred to as the Clarke Orbit (named in honor of Arthur C. Clarke, a science fiction writer who first postulated the characteristics of this orbit in 1945).

- **geocaching**

  A high-tech version of hide-and-seek. Geocachers seek out hidden treasures utilizing GPS coordinates posted on the Internet by those hiding the cache; The sport where you are the search engine, and you have to hunt for treasure caches with your GPS; A high-tech treasure hunt using GPS to find hidden caches.

- **geocenter**

  The center of the earth's mass.

- **geodashing**

  An outdoor sport in which players use GPS receivers on a playing field that covers the entire planet. The waypoints, or dash points, to be reached are randomly selected. The win goes to the team that reaches the most dashpoints in a given time.

- **geodesy**

  The branch of applied mathematics that deals with the measurement, curvature, and shape of the earth; The science related to the determination of the size and shape of the Earth by direct measurements.

- **geodetic**

  Of, pertaining to, determined, or effected by geodesy; Geographic, or pertaining to geodesy. Latitude and longitude readings are geodetic coordinates.

- **geodetic coordinates**

  A coordinate system whose elements are latitude, longitude and geodetic height. The latitude is an angle based on the perpendicular to the ellipsoid. Longitude is the angle measured in the XY plane.

- **geodetic datum**

  A math model representing the size and shape of the earth (or a portion of it); The reference ellipsoid surface that defines a coordinate system. Picked to try and best fit part or all of the geoid; A mathematical model designed to best fit part or all of the geoid. It is defined by an ellipsoid and the relationship between the ellipsoid and a point on the topographic surface established as the origin of datum. This relationship can be defined by six quantities, generally (but not necessarily) the geodetic latitude, longitude, and the height of the origin, the two components of the deflection of the vertical at the origin, and the geodetic azimuth of a line from the origin to some other point. The Global Positioning System uses WGS-84; A specifically oriented reference ellipsoid. Typically, eight parameters are required to define a geodetic datum: two to specify the dimensions of the ellipsoid, three to specify the location of its center with respect to the earth's center of mass, and three to specify the orientation of the ellipsoid with respect to the average spin axis of the earth and the Greenwich reference meridian.

- **geodetic datum (horizontal datum)**

A specifically oriented ellipsoid typically defined by eight parameters which establish its dimensions, define its center with respect to Earth's center of mass and its orientation in relation to the Earth's average spin axis and Greenwich reference meridian.

- **geodetic height**

The height of a point above an ellipsoidal reference surface. Also known as ellipsoidal height. The difference between the geodetic height of a point and its orthometric height is equal to the geoidal height.

- **geodetic height (ellipsoidal height)**

The geodetic height is the distance of a point above an ellipsoidal surface. The difference between a point's geodetic height and its orthometric height equals the geoidal height.

- **geodetic survey**

Global surveys for the establishment of control networks (comprised of Reference or Control Points), which are the basis for accurate land mapping. May be carried out using either terrestrial or satellite positioning (e.g. GPS) techniques. The outcome is a network of benchmarks which are a physical realisation of the Geodetic Datum or Reference System; Global surveys done to establish control networks (comprised of reference or control points) as a basis for accurate land mapping.

- **geofence**

An imaginary boundary set which contains a GPS device-if that GPS device exceeds the boundary, an alarm is activated.

- **geographic information system (gis)**

A computer system or software capable of assembling, storing, manipulating, and displaying geographically referenced information (i.e., data identified according to their location). In practical use, GIS often refers to the computer system, software, and the data collection equipment, personnel, and actual data; A computer-based system that is capable of collecting, managing and analysing geospatial data. This capability includes storing and utilising maps, displaying the results of data queries and conducting spatial analysis.

- **geoid**

The fundamental surface in Geodesy. It is defined as the equipotential surface of the gravity field that most closely approximates the Mean Sea Level. (The MSL deviates from the Geoid surface by 1-2 metres due to the Sea Surface Topography caused by wind-driven or geostrophic currents.) The Geoid is the Vertical Datum surface both from a mathematical viewpoint (i.e., the sum of the Orthometric Height and the Geoid Height equals the Ellipsoidal Height of a point), as well as in practice by making the land height system synonymous with "height above MSL". Models of the Geoid Height have been determined from the combined processing of satellite-derived potential models, surface gravity observations and the ocean gravity anomalies derived from Satellite Altimetry. Their accuracy may range from a few metres in the open ocean areas, down to the few decimetre level in land areas where there is a good coverage of surface gravity; That equipotential surface (a surface of equal gravity potential) which most closely matches mean sea level. An equipotential surface is normal to the gravity vector at every point; Representation of the earth's surface considered as a sea level surface extended continuously through the continents. The actual geoid is an equipotential surface coincident with mean sea level to which at every point the plumb line (direction of gravitational pull) is perpendicular; The particular equipotential surface which coincides with mean sea level, and which may be imagined to extend through the continents. This surface is everywhere perpendicular to the force of gravity.

- **geoidal height (geoidal separation; undulation)**

The height of a point on the geoid above the ellipsoid measured along a perpendicular to the ellipsoid. Height above the geoid, often referred to as elevation above mean sea level.

- **geomatics**

The science and technology of gathering, analyzing, interpreting, distributing, and using geographic information

- **geometric dilution of precision**

Since a GPS receiver determines position by triangulation, when the GPS satellites that are being received are clustered too close together, the positional accuracy determined by the receiver is diluted. The wider the angle between satellites, the better the measurement.

- **geophysics**
  the mapping of subterranean soil characteristics using non-invasive techniques such as sound.

- **geoscience**
  a range of scientific disciplines (geology, geophysics, petroleum engineering, bio stratification, geochemistry, etc.) related to the study of rocks, fossils and fluids.

- **geostationary**
  Satellite in orbit along the equator that results in a constant fixed position over a particular reference point on the earth's surface; A satellite in a position always fixed to a relative point on the earth; Refers to a geosynchronous satellite angle with zero inclination. so the satellite appears to hover over one spot on the earth's equator.

- **geostationary orbit**
  A satellite orbit at such an altitude (approximately 22,300 miles) over the equator that the satellite appears to be fixed above a given point.

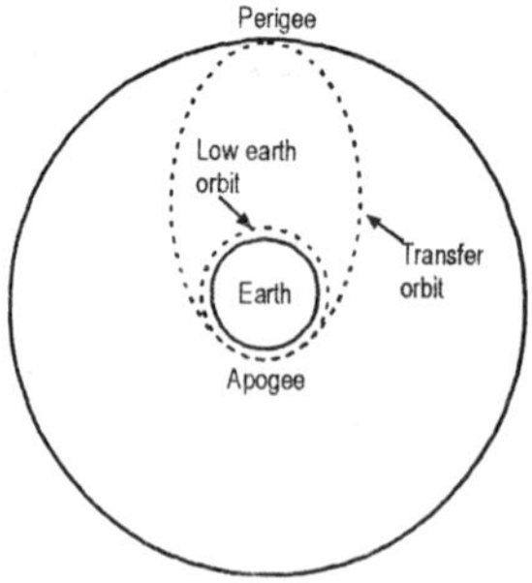

Final geostatonary orbit

- **geostationary satellites**
  Those satellites situated in a constant orbit position relative to a given area of the globe with the purpose of maintaining constant coverage of that area.

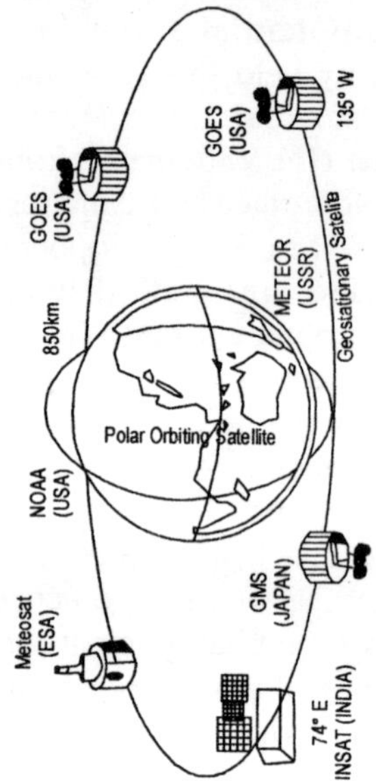

- **geosynchronous**
  The Clarke circular orbit above the equator. For a planet the size and mass of the earth, this point is 22,237 miles above the surface.

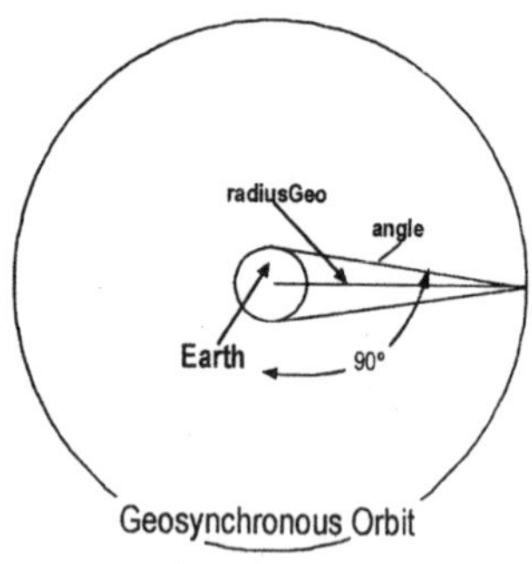

- **geosynchronous orbit**

  A specific orbit around where a satellite rotates around the earth at the same rotational speed as the earth. A satellite rotating in geosynchronous orbit appears to remain stationary when viewed from a point on or near the equator. It is also referred to as a geostationary orbit; Satellite in geosynchronous orbit travel around the Earth in an area approximately 22,300 miles from the surface at the same rate that the Earth turns, therefore completing one revolution every 24 hours.

- **geotechnics**

  the determination of subterranean soil characteristics using invasive techniques such as probing, drilling and sampling.

- **getting**

  Ivan A. Getting, Father of the GPS system born 1912. Bradford Parkinson worked on the implemented the system.

- **gigahertz (ghz)**

  One billion cycles per second. Signals operating above 3 Gigahertz are known as microwaves. above 30 GHz they are know as millimeter waves. As one moves above the millimeter waves signals begin to take on the characteristics of light waves.

- **gis**

  Geographic Information System. A computer based system that is capable of collecting, managing and analyzing geographic spatial data; A category of computer programs and applications that are used to organize, analyze and display geographic information.

- **gll**

  A data sentence that is part of the NMEA protocol for transferring information between electronic devices. The GLL sentence contains position information.

- **global beam**

  An antenna down-link pattern used by the Intelsat satellites, which effectively covers one-third of the globe. Global beams are aimed at the center of the Atlantic, Pacific and Indian Oceans by the respective Intelsat satellites, enabling all nations on each side of the ocean to receive the signal. Because they transmit to such a wide area, global beam transponders have significantly lower EIRP outputs at the surface of the Earth as compared to a US domestic satellite system which covers just the continental United States. Therefore, earth stations receiving global beam signals need antennas much larger in size (typically 10 meters and above (i.e.30 feet and up).

- **global navigation satellite system (glonass)**

  This is the Russian counterpart to GPS. GLONASS provides worldwide coverage, however, its accuracy performance is optimized for the northern latitudes and is specified as identical to that of GPS SPS; This is the Russian counterpart to the NAVSTAR system run by the United States. It also offers global positioning capability although performance is optimized for use in the northern hemisphere, where performance is comparable to that of NAVSTAR. Some GPS receivers use a combination of both NAVSTAR and GLONASS to provide enhanced capabilities.

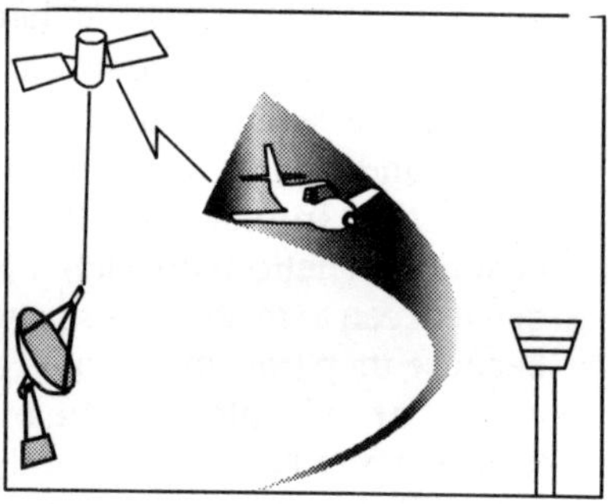

This is an umbrella term used to describe a generic satellite-based navigation/positioning system. It was coined by international agencies such as the International Civil Aviation Organisation (ICAO) to refer to both GPS and GLONASS, as well as any augmentations to these systems, and to any future civilian developed satellite system. For example, the Europeans refer to GNSS-1 as being the combination of GPS and GLONASS, but GNSS-2 is the blueprint for an entirely new system.

- **global orbiting navigation satellite system (glonass)**

This is the Russian counterpart to GPS. It consists of a constellation of 24 satellites (though the number may vary due to difficulties in funding for the system) transmitting on a variety of frequencies in the ranges from 1597-1617MHz and 1240-1260MHz (each satellite transmits on two different L1 and L2 frequencies). GLONASS provides worldwide coverage, however, its accuracy performance is optimised for northern latitudes, where it is better than GPS's SPS (there being no "Selective Availability" on GLONASS satellites). GLONASS positions are referred to a different Datum to those of GPS, i.e. PZ90 rather than WGS84.

- **global positioning system (gps)**

A system for providing precise location which is based on data transmitted from a constellation of 24 satellites; A global navigation system based on 24 or more satellites orbiting the earth at an altitude of 12,000 statue miles and providing very precise, worldwide positioning and navigation information 24 hours a day, in any weather. Also called the NAVSTAR system. For more information, see About GPS; A system for providing precise location which is based on data transmitted from a constellation of 24 satellites. It comprises three segments:

(a) the Control Segment,

(b) the Space Segment, and

(c) the User Segment.

The NAVSTAR Global Positioning System consists of a space segment (24 orbiting satellites, four equally spaced around each of six different orbital planes), the control segment (five monitor stations, one master control station and three uplinks). The GPS receiver is referred to as the user segment. NAVSTAR satellites carry atomic clocks and broadcast coherent signals so that the position to each satellite may be determined by calculating Doppler shift. See also GLONASS. A global navigation system that is based on triangulation from a constellation of 24 satellites orbiting the earth. A GPS receiver pinpoints its position on earth by measuring its distance from the satellites. It does so by calculating the time it takes for a coded radio message to pass from the satellite to the GPS unit. A GPS unit needs at least three measurements to determine its exact position.

- **globalstar**

Globalstar differs from Iridium in several important ways:

1. Globalstar orbits have an inclination of 52 degrees, compared to the near-polar 86.4-degree orbits used by Iridium. Globalstar also does not cover polar areas due to the lower orbital inclination.
2. The Globalstar system uses the Qualcomm CDMA air interface; however, handsets use standard GSM SIMs.

- **glonass**

The Global Orbiting Navigational Satellite System; the Russian counterpart to the United States' GPS system.

- **gmat**

Greenwich Mean Astronomical Time. Prior to 1925, in astronomical and nautical almanacs, a day of Greenwich Mean Time began at noon. This reckoning of Greenwich Mean Time is now called Greenwich Mean Astronomical Time, and is no longer used. Persons using old editions of the almanacs for historical research should be aware of the previous convention.

- **gmt (greenwich meridian time)**

The time zone which includes Greenwich, England, and which is bisected by 0 Degree Longitude This is the time notation which is used for booking international satellite time.

- **gnis-geographic names information system**

Developed by the USGS in cooperation with the U.S. Board on Geographic Names (BGN). It contains information about almost 2 million physical and cultural geographic features in the United States.

- **good gdop**

GDOP is computed from the geometric relationships between the receiver position and the positions of the satellites the receiver is using for navigation. For planning purposes GDOP is often computed from Almanacs and an estimated position.

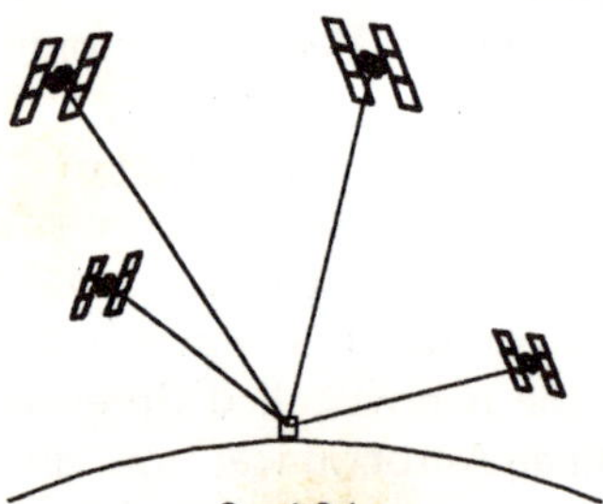

Good Gdop

- **good gdop bad visibility**

GDOP terms are usually computed using parameters from the navigation solution process.

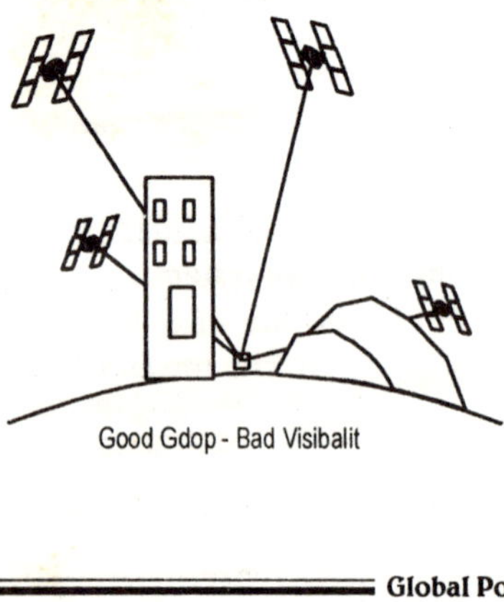

Good Gdop - Bad Visibalit

- **goto**

A route consisting of one leg, with your present position being the start of the route and a single defined waypoint as the destination.

- **gprs (general packet radio service)**

Standard for data transmission by packet, with a speed of 64 Kbits/s. gprs-general packet radio service technology that enables high-speed wireless Internet and other data communications for uses such as mobile phones and computers.

- **gps**

Global Positioning System, consisting of: - a space segment (24 NAVSTAR satellites in 6 different orbits) - the control segment (5 monitor stations, 1 master control station and 3 upload stations) - the user segment (GPS receivers) - NAVSTAR satellites carry extremely accurate atomic clocks and broadcast coherent simultaneous signals; GPS, which stands for Global Positioning System, is a navigational system designed and operated by the U. S. Department of Defense (DOD). Coded satellite signals from three or more GPS satellites are processed by your vehicle's navigation computer every second, which then calculates the vehicle's velocity and geographical location; A system of 24 satellites used to locate any point on the earth by triangulation and distance measuring.

- **gps (system) time**

The time scale to which GPS signals are referenced. GPS Time derives from a composite or clock consisting of all operational monitor station and satellite atomic clocks. It is steered over the long run to keep it within about 1 micro-second of UTC, as maintained by the Master Clock at the U.S. Naval Observatory, ignoring the UTC leap seconds. At the integer second level, GPS Time equaled UTC in 1980, but currently, due to the leap seconds that have been inserted into UTC, it is ahead of UTC by 10 seconds.

- **gps applications, costs, and signals**

Military: The primary military purposes are to allow improved command and control of forces through improved locational awareness, and to facilitate accurate targeting of smart bombs, cruise missiles, or other munitions.

Civilian Navigation: This Taxi in Kyoto, equipped with GPS navigation, is an example of how GPS technology can improve everyday life. Even fixed systems may use GPS, in order to get precise time.

GPS ACCURACIES, COSTS, AND SIGNALS

GPS Approach	Accuracy Estimate	Receiver Cost Estimate	GPS Signals				
			L1 C/A Code	L1 P-Code	L1 Carrier	L2 P-Code	L2 Y-Code
SPS Navigation	100M	$1,000	×				
SPS Differential >30KM	10M	#5,000	×				
SPS Differential <30KM	1M	$5,000	×				
PPS Navigation	10M	$10,000	×	×		×	
Anti-Spoofing Navigation	10M	$20,000?	×	×	×	×	×
L1 Carrier Phase Survey	0.1M	$10,000	×		×		
L1L2 Carrier Phase Survey	0.01	$15,000	×	×	×	×	

- **gps axis**

A satellite based navigation system providing accuracy usable for side scan sonar surveys on a worldwide basis. GPS has become a universal, reliable positioning system. Inherent errors in GPS (implemented by the Department of Defense) create inaccuracies of more than 100 meters. Differential base stations can reduce these errors to less than two meters but are time consuming to survey-in and have limited range. Still, GPS is a convenient navigation system for sonar survey operations.

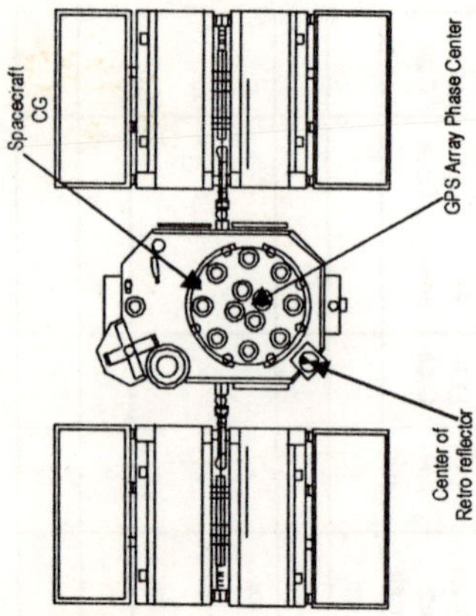

- **gps c/a code generator**

Three binary codes shift the L1 and/or L2 carrier phase. The C/A Code (Coarse Acquisition) modulates the L1 carrier phase. The C/A code is a repeating 1 MHz Pseudo Random Noise (PRN) Code.

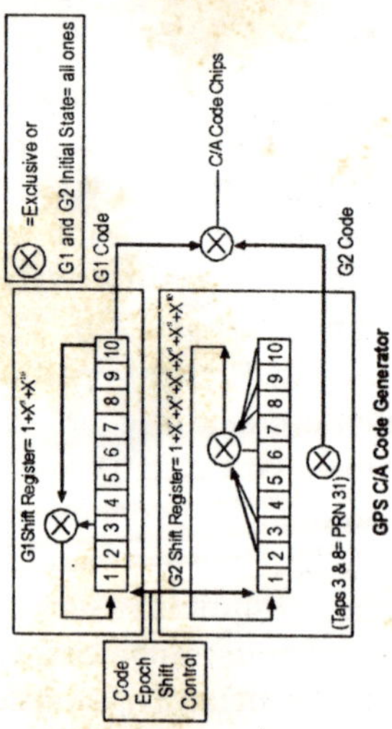

- **gps carrier-phase positioning**

In standard differential GPS, clock and ephemeris errors can be reduced for the Carrier Phase method by using two receivers. One receiver is located at a base station with a known position (benchmark).

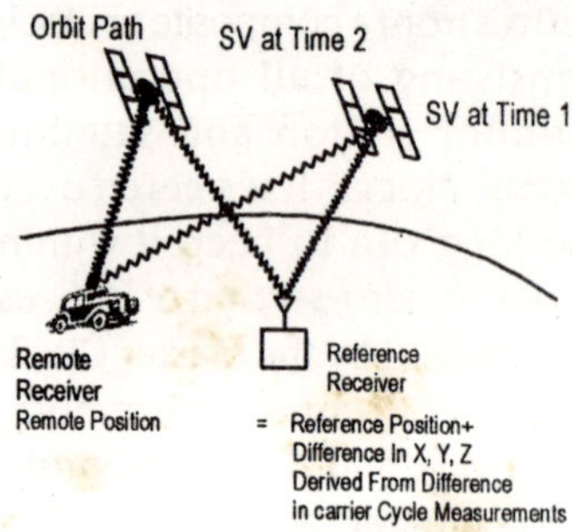

GPS Carrier Phase Positioning

- **gps common view time transfer**

In a vehicle tracking system computers aboard moving vehicles determine periodically their geographical position and transmit it to a control center.

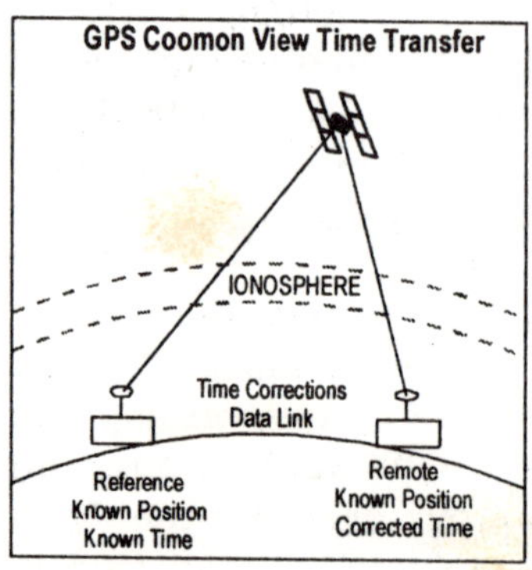

• **gps constellation**

GPS constellation neared completion in the early 1990s, domestic manufacturers were well aware of the commercial potential of GPS.

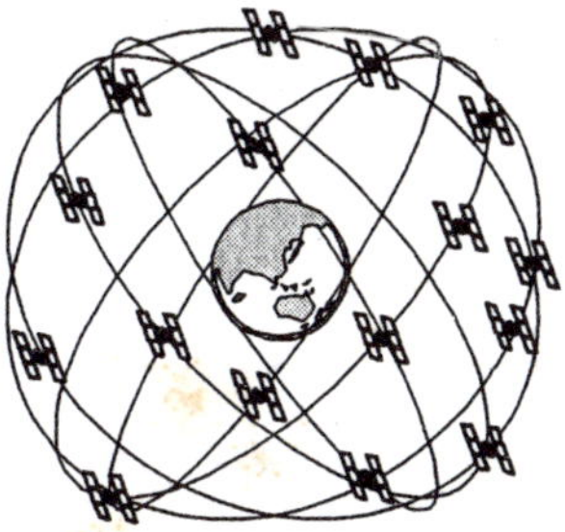

GPS Nominal Constellation
24 Satellites in 6 Orbital Planes
4 Satellites in each Plane
20,200 km Altitudes, 55 Degree Inclination

• **gps control monitor**

The passive monitor stations are GPS receivers which track all satellites in their range and collect data of the satellite signals. The raw data are then sent to the mater control station where the data are processed.

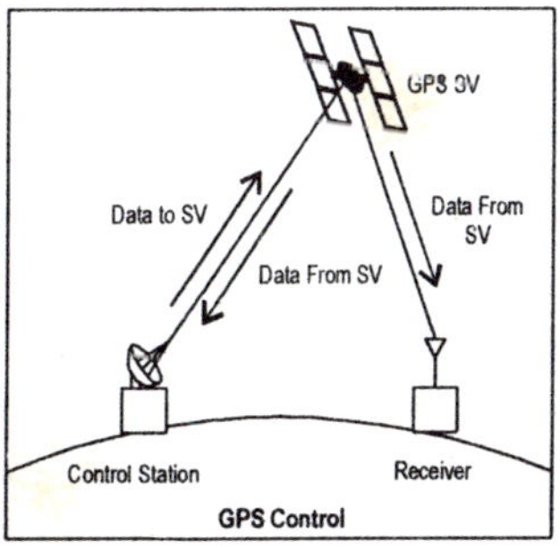

GPS Control

• **gps error technique**

Ideally, GPS receivers would easily be able to convert the C/A and P(Y)-code measurements into accurate positions. However, a system with such complexity leaves many openings for errors to affect the measurements.

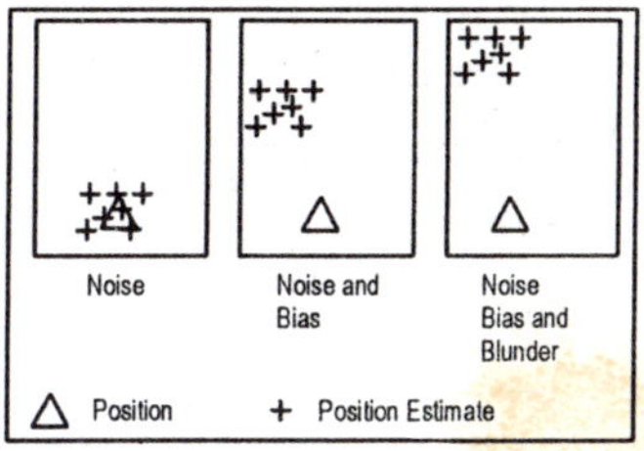

• **gps icd-200**

Specification that contains full technical details on how a GPS receiver must interface with the system.

• **gps navigation**

By 2000, in-dash GPS-based "navigation systems" were standard or at least an option in luxury cars, and third-party systems became available for all cars.

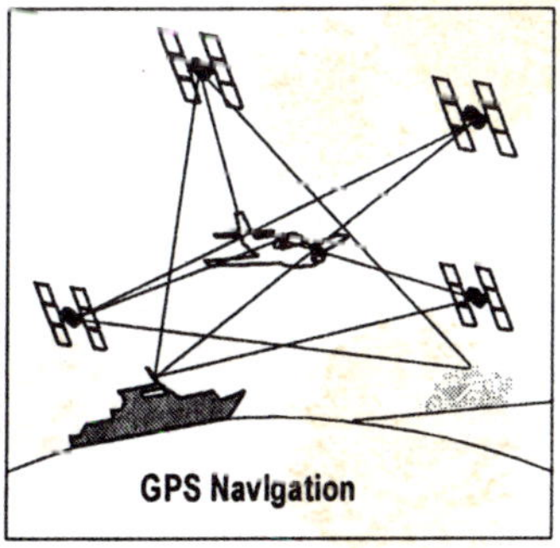
GPS Navigation

• gps navigation data format

- Latitudes are parallel to the Equator and range from -90 degrees (South Pole) to+90 degrees (North Pole).
- The northern hemisphere has positive latitudes.

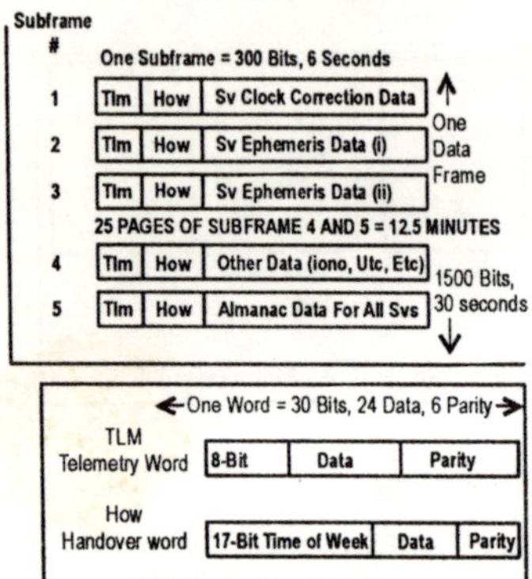

GPS Navigation Data Format

• gps nominal orbit planes

The complete nominal GLONASS constellation consists of 24 satellites, 21 operating and three on-orbit 'spares' placed in three orbital planes. Each plane contains eight satellites identified by "slot" number, which defines the corresponding orbital plane and the location within the plane.

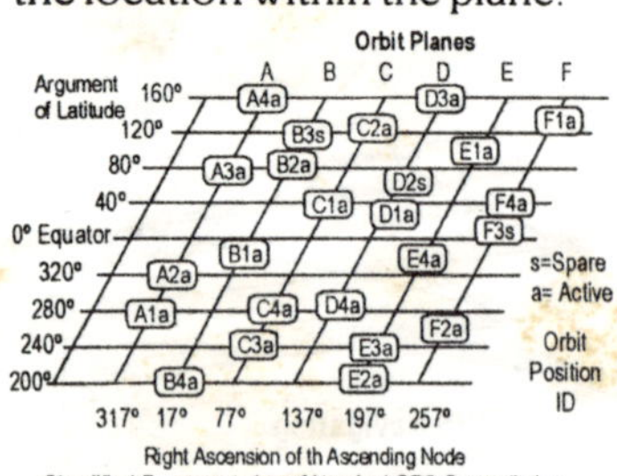

Simplified Representation of Nominal GPS Constellation

• gps satellite signals

Satellite navigation systems can be used for determining a precise location and providing a highly accurate time reference almost anywhere on Earth or in Earth orbit.

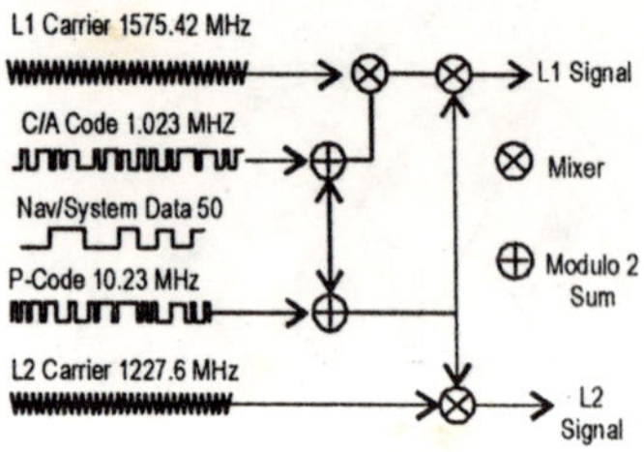

GPS Satellite Signals

• gps signals

In essence, the GPS system is made of a constellation of 24 satellites orbiting at an average altitude of 20,000 kms, scattered on 6 orbital planes equaly spaced. As a result, at least three satellites are always visible at anytime and from anyplace on earth.

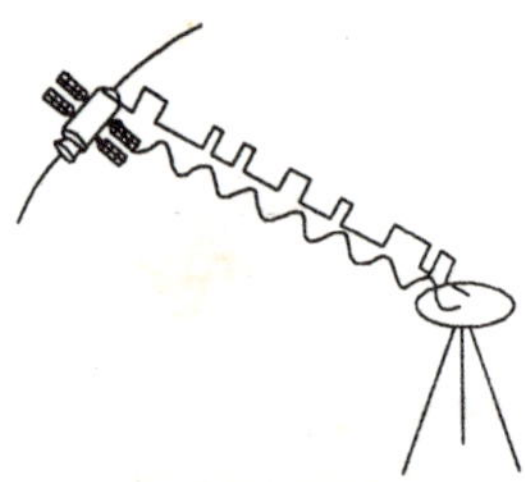

Each satellite emits a coded signal which contains essential informations like its position and the exact timing of the signal emission to earth. Therefore, nothing but a simple receiver is needed to measure the elapsed time between emission and reception of the signal. The satellite to station distance is simply deduced from this travel time. Three different measurements made on three different satellites give the three distances needed to determine the three coordinates of the station position: latitude, longitude, and altitude. This type of measurements is known as "pseudo-range" measurements in GPS jargon. Every satellite emits two types of pseudo-ranges : A precise code (P code) which enables a position precision of around 10 meters and a coarse code (C/A code) which allows a precison of around 100 meters. The precise code is encrypted to deny the precise positioning to anyone else than the American military. Therefore, the only precision available to civilian applications based on the pseudo-ranges measurements is this of the C/A code. The precision achieved by pseudo-range measurements is not good enough for most of the GPS applications in geophysics. For plate tectonic for exemple, it is required that measurements be precise at the centimeter level (even millimeter), if one is to be able to detect motions of a few centimeters per year (or less) without having to wait. Again, to prevent potential "hostiles" to be able to get precise positions from GPS, the system was built with the capacity to degrade its precison. This is achieved by artificial degradation of the precision with which a certain number of parameters are broadcasted by the satellites. For any of those numbers, the last "byte" which contains the final digits (and therefore the full precision) is deliberately jammed, that is single bit positions are exchanged using an unknown algorithm. This dithering, known as Selective Availability, affects first the clock of the satellites which gives the time tag for signal emission, and second the satellite orbits which give the position of the emission point. In practical, it is possible to go around this problem by using differential GPS. This technique consists in using a reference station which position is known with accurate precision. At every moment, the difference between the wrong measured position and the true known position is radio broadcasted as a correction valid in the whole area of the reference station. for hundred of years to generate sizeable displacements. Then, a different technique is applied. It consists in acquiring measurements of the satellite-station distance directly on the carrier wave of the GPS signal. In principal it is similar to pseudo-range measurements, except that the wavelength (or characteristic size) of the signal is considerably reduced, which allows a far better precision.

- **gps surveying**

Conventional static GPS surveying has the following characteristics:

(1) The points being coordinated are not moving, i.e. they are "static".

(2) GPS data are collected over some "observation session", typically ranging in length from an hour to several hours (or perhaps days for very precise GPS Geodesy applications).

(3) The results are not required immediately, for in-the-field use.

(4) The relative positioning mode of operation is the only mode employed, requiring the use of a minimum of two GPS receivers for all survey work.

(5) The measurements used for data reduction are those made on the transmitted L-Band carrier wave, requiring specialised hardware and software.

(6) A variety of processing algorithms can be employed, including "bias-free" and "bias-fixed" solutions.

(7) Mostly associated with the traditional surveying and mapping functions.

Since the late 1980's considerable attention has been paid to the first three points, as they were considered to be unnecessarily restrictive for typical GPS surveying applications. As a result of vigorous R&D, new GPS surveying methodologies have been developed, which complement the "conventional static" technique. These modern GPS Surveying techniques are given a variety of names but the following are considered generic:

(a) rapid static positioning techniques,

(b) "stop & go" techniques, and

(c) "on-the-fly" positioning techniques.

Each of the techniques represents a technological solution to the problem of obtaining high productivity (measure as many baselines in as short a period of time as possible) and/or versatility (for example, the ability to obtain results even while the receiver is in motion) without sacrificing very much in terms of accuracy and reliability. None of these techniques is as accurate or reliable as conventional static GPS surveying, and each of these techniques has its special strengths and weaknesses. They represent the state-of-the-art in precision carrier phase-based GPS positioning, and are a direct outcome of considerable innovation by instrument manufacturers seeking to address surveying and non-surveying applications.

- **gps sv and receiver xyz**

Receiver position is computed from the SV positions, the measured pseudo-ranges (corrected for SV clock offsets, ionospheric delays, and relativistic effects), and a receiver position estimate (usually the last computed receiver position).

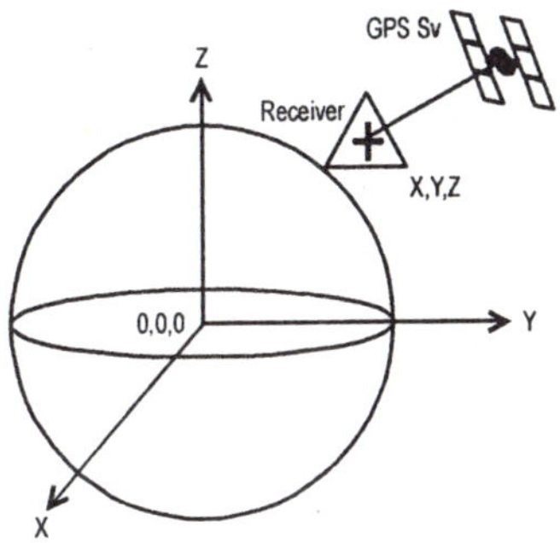

Earth Centered, Earth Fixed X,Y,Z

- **gps system time**

The time scale to which GPS signals are referenced. GPS Time derives from a composite or paper clock consisting of all operational monitor station and satellite atomic clocks. It is steered over the long run to keep it within about 1 micro-second of UTC, as maintained by the Master Clock at the U.S. Naval Observatory, ignoring the UTC leap seconds second.

- **gps technology**

GPS was developed and is operated by the U.S. Department of Defense. It was originally called NAVSTAR. GPS has applications beyond navigation and location determination. GPS can be used for cartography, forestry, mineral exploration, wildlife habitation management, monitoring the movement of people and things and bringing precise timing to the world.

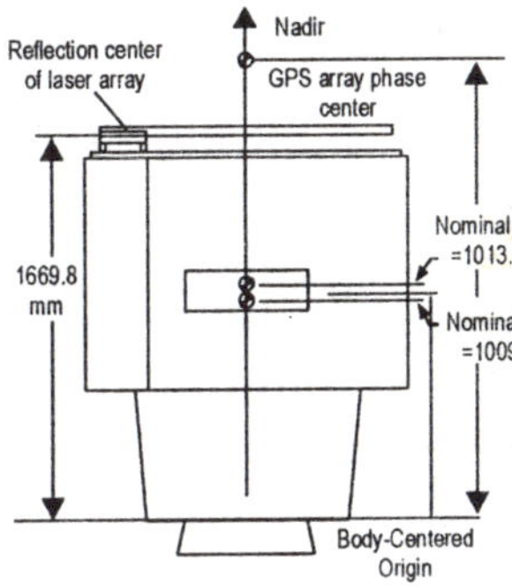

- gps time (gpst)

GPST is a form of Atomic Time, as is, for example, Coordinated Universal Time (UTC). GPST is "steered" over the long run to keep within one microsecond of UTC. The major difference is that while "leap seconds" are inserted into the UTC time scale every 18 months or so to keep UTC approximately synchronised with the earth's rotational period (with respect to the sun), GPST has no leap seconds. At the integer second level, GPST matched UTC in 1980, but because of the leap seconds inserted since then, GPST is now (end 1998) ahead of UTC by 12 seconds (plus a fraction of a microsecond that varies from day to day). The relationship between GPST and UTC is transmitted within the Navigation Message.

- **gps week**
  Incremental number of weeks, starting at 0 hour UTC on the date January 6, 1980. April 6, 1997 is the first day of GPS week 900; GPS time started at Saturday/Sunday midnight, January 6, 1980. The GPS week is the number of whole weeks since GPS time zero.

- **gpx**
  GPS exchange format is an XML scheme designed for transferring GPS data between software applications. It can be used to describe waypoints, tracks, and routes.

- **graphical user interface (gui)**
  An onscreen system (including menus, maps, etc.) to control a mobile audio/video system. The user employs touch controls or cursors on the monitor to navigate through system functions and parameters.

- **graticule**
  A network of parallels and meridians on a map or chart.

- **gravity**
  Gravity is the mutual force of attraction between all particles or bodies that have mass. Gravity often refers to the vector sum of the gravitational attraction of various masses within the planet plus the centrifugal force caused by the rotation of the Earth.

- **gravity void**
  A block or area of blocks within the gravity measurement database without observations. A geoid model relying upon this database would be weak and possibly in error at these blocks.

- **grayline®**
  This patented Lowrance feature helps you distinguish between hard and soft bottoms, where the thicker the gray band, the harder the bottom. GRAYLINE® also helps separate fish and important structures on or near the bottom from the actual bottom. Since more active-feeding fish hold close to hard bottoms and structures, GRAYLINE® helps you find more potentially productive water quickly.

- **great circle**
  The intersection of a plane through the center of the Earth and the surface of the Earth. The shortest distance between two points on the Earth is a great circle route. All longitudes are great circles. The only latitude that is a great circle is the Equator; A circle described by the intersection of the surface of the Earth with a plane passing through the center of the Earth. The shortest distance between two points on the surface

of the Earth is a segment of a great circle. All longitudes are great circles; the only latitude that is a great circle is the equator.

- **greenwich mean time**
  (GMT) See Universal Time. In discussions about GPS, they are often used interchangeably; The mean solar time for Greenwich, England, which is located on the Prime Meridian (zero longitude). Based on the rotation of the earth, GMT is used as the basis for calculating standard time throughout most of the world.

- **grid**
  Also "Coordinate System". A coordinate system that projects the earth on a flat surface, using square zones for position measurements; A pattern of regularly spaced horizontal and vertical lines forming square zones on a map used as a reference for establishing points. Grid examples are UTM, MGRS, and Maidenhead; A map coordinate system that projects the surface of the earth onto a flat surface such as a "map", using square zones for position measurements. Common map grids include that defined by the UTM (Universal Transverse Mercator) projection.

- **grid north**
  The direction that north/south lines of a grid point.

- **grid zone**
  One of the 60 zones covered by the Universal Transverse Mercator (UTM) grid system of navigation.

- **ground plane**
  A large flat metal surface, or electrically charged field, surrounding a GPS antenna used to deflect errant signals (multipath) reflected from the ground and other near-by objects.

- **ground plane**
  A large flat metal surface, or electrically charged field, surrounding a GPS antenna used to deflect errant signals (multipath) reflected from the ground and other near-by objects.

- **ground segment**
  A term which describes that portion of the total communications satellite system which is situated on the earth such as transmitting and receiving antenna, RF (radio frequency) signal generating facilities, RF receivers and digital encoding equipment and decoders.

- **ground speed**
  The velocity you are traveling relative to a ground position. Typically measured in "knots" (nautical miles per hour), but may be expressed in km/hr or m/s.

- **ground station**
  A radio station, on or near the surface of the Earth, designed to

transmit or receive to/from a spacecraft.

- **ground wave**
  A radiowave that travels along the earth's surface.

- **groundtrack**
  The imaginary line traced on the surface of the Earth by the sub-satellite point (SSP).

- **grs 80**
  The reference ellipsoid of the NAD83 coordinate system.
  a = 6378137.000 m,
  b = 6356752.314 m,
  1/f = 298.257 222 101

- **gsm (global system for mobile communications)**
  Digital cellular telephony system used in ITS services such as traffic information, emergency call and fleet management.

- **guard band**
  Referring to the frequency that is left vacant between two channels. The margin of safety that is provided helps to insure that adjacent signals do not interfere with one another.

- **guard channel**
  Television channels are separated in the frequency spectrum by spacing them several megahertz apart. This unused space serves to prevent the adjacent television channels from interfering with each other.

- **gwen**
  Air Force Ground Wave Emergency Network

- **gyroscope**
  Navigation systems use an internal gyroscope to track the turns a vehicle makes, which helps the system determine the direction in which a vehicle is traveling.

- **hachure**
  Short lines attached to, and perpendicular with a contour line usually used to indicate a depression contour.

- **half transponder**
  A method of transmitting two TV signals through a single transponder through the reduction of each TV signal's deviation and power level. Half-transponder TV carriers each operate typically 4 dB to 7 dB below single-carrier saturation power.

- **half-duplex operation**
  That mode of operation in which communication between two terminals occurs in either direction but in only one direction at a time. Contrast with duplex or simplex operation. Note: Half-duplex operation may occur on half-duplex circuits or on duplex circuits, but it may not occur on simplex circuits.

- **hand held gps unit**

A handheld GPS is the best choice for outdoor pursuits, since their small screens are more easily read at arm's length than in a moving car.

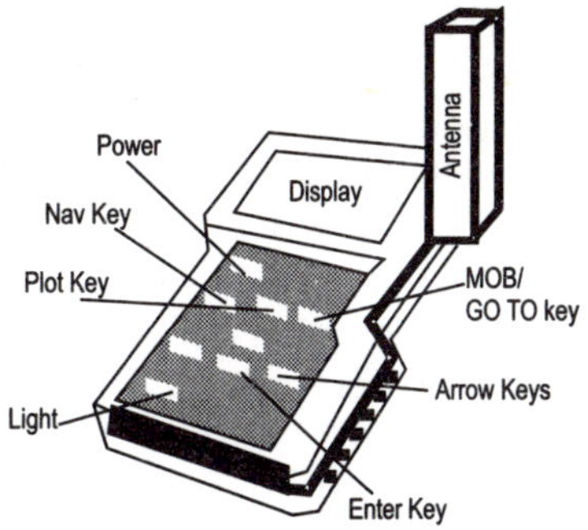

- **hand-over word (how)**

The second word in each subframe of the navigation message. It contains the Z-count at the leading edge of the next subframe and is used by a GPS receiver to determine where in its generated P-code to start the correlation search process; The word in the GPS message that contains synchronization information for the transfer of tracking from the C/A to P code.

- **hardover word**

The word in the GPS message that contains synchronization information for the transfer of tracking from the C/A to P code.

- **hdg heading**

The direction you are facing, defined as an angle from North.

- **hdop**

HDOP = Horizontal Dilution of Precision (Latitude, Longitude).

- **hdtv (high definition television)**

The term refers to the new format of digital signal to be seen on the television screen. The HDTV format will be delivered by the ATV (Advanced Television) digital system developed by the members of the Grand Alliance (AT&T, David Sarnoff Research Center, General Instrument, Massachusetts Institute of Technology, North American Philips, Thomson Consumer Electronics and Zenith Electronics). This new system will provide a television picture with about 1000 scanlines per picture (compared to 525 for existing standard) and a picture aspect ratio of 16:9 (compared to 4:3 for existing standard). This format will be broadcast over the UHF spectrum.

- **headend**

Electronic control center - generally located at the antenna site of a CATV system - usually including antennas, preamplifiers, frequency converters, demodulators and other related equipment which amplify, filter and convert incoming broadcast TV signals to cable system channels.

- **heading**

The direction in which a vehicle is moving. For air and sea

operations, this may differ from actual Course Over Ground (COG) due to winds, currents, etc.

- **healthy**

A term used when an orbiting GPS satellite is suitable for use. "State" is also used to refer to satellite health.

- **height (ellipsoidal)**

The height coordinate determined from GPS observations is related to the surface of a Reference Ellipsoid. The coordinates are derived initially in the 3-D Cartesian system (as XYZ values), and then for display/output purposes they are transformed to Latitude, Longitude and (Ellipsoidal) Height using well known formulae to an ellipsoid such as that associated with the WGS84 Datum (semi-major axis: 6378137m; inverse flattening: 298.257223563). The surface of the ellipsoid is the zero ellipsoidal height datum. In Relative Positioning, the height component of the receiver whose coordinates are being determined relative to the Base Station can also be related to an ellipsoid by transforming the baseline vector from the 3-D form (DXDYDZ) to a change in Latitude, change in Longitude, and change in Ellipsoidal Height.

- **height (orthometric)**

The Orthometric Height is the height of a station on the earth's surface, measured along the local plumbline direction through that station, above the Geoid surface. It is approximated by the "Height Above Mean Sea Level", where the MSL Datum is assumed to be defined by the mean tide gauge observations over several years. The relationship between Orthometric Height (H) and Ellipsoidal Height (h) is :

$$h = H + N,$$

where N is the Geoid Height or Geoid Undulation with respect to the Reference Ellipsoid. Orthometric Height is traditionally derived from geodetic levelling (using such techniques as optical levelling, trigonometrical levelling, barometric levelling).

- **height aiding**

An accuracy enhancement technique in which the known height of the receiver is entered into the navigation solution, which in effect provides another satellite's range.

- **hertz**

A unit used to measure a wave's frequency, one cycle per second; The name given to the basic measure of radio frequency characteristics.

An electromagnetic wave completes a full oscillation from its positive to its negative pole and back again in what is known as a cycle. A single Hertz is thus equal to one cycle per second; Cycle per second; a measure of electromagnetic frequency that represents the number of complete electrical waves in a second. One kiloHertz (kHz) is one thousand cycles per second; one megaHertz (MHz) is one million; one gigaHertz (gHz) is one billion. hertz (hz): a basic measurement of the frequency of an electrical signal equal to one cycle per second.

1. kilohertz (khz): one thousand cycles/second (10p3)
2. megahertz (mhz): one million cycles/second (10p6)
3. gigahertz (ghz): one billion cycles/second (10p9)
4. terahertz (thz): one trillion cycles/second (10p12)
5. petahertz (phz): one quadrillion cycles/second (10p15)
6. exahertz (ehz): one quintillion cycles/second (10p18).

An electromagnetic wave oscillates from positive to negative poles. an oscillation from positive to negative and back to positive again is one complete cycle.

- **high definition television**

An improved television system which provides approximately twice the vertical and horizontal resolution of existing television standards. It also provides audio quality approaching that of compact discs.

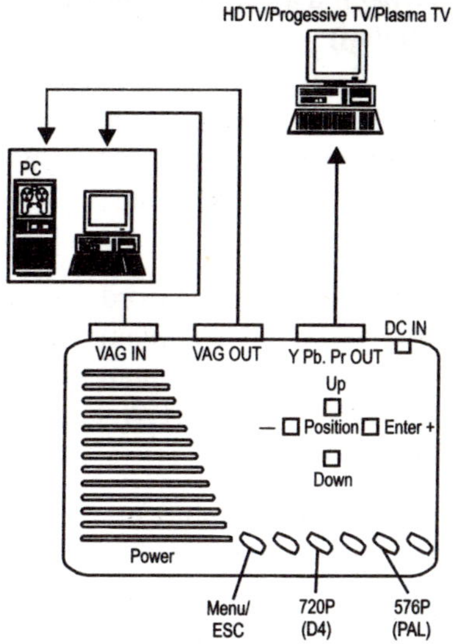

- **high frequency (hf)**

Radio frequencies in the band from 3 to 30 MHz.

- **high power amplifier (hpa)**

Equipment used to amplify signals.

- **high-power satellite**

A satellite with greater than 100 Watts of transponder radio frequency (RF) transmitting power.

- **hmi (human machine interface)**

Previously known as man-machine interface (MMI); means by which a user interacts with a machine and includes simple and advanced functions such as voice recognition, speech synthesis and touch screens.

- **horizontal dilution of precision (hdop)**
  The effects on accuracy of the combined errors in a two-dimensional fix obtained from crossing two lines of position.
- **how**
  Handover word. The word in the GPS message that contains time synchronization information for the transfer from the C/A code to the P-code.
- **hp (high-performance)**
  decimetre positioning accuracy.
- **htdp**
  Horizontal Time Dependent Positioning model A computer database and interpolation program developed by NGS to predict horizontal displacements between coordinate points over time. The program can work backwards in time where it includes earthquake parameters, or forward in time where only the secular motion is analyzed.
- **hub**
  The master station through which all communications to, from and between micro terminals must flow. in the future satellites with on-board processing will allow hubs to be eliminated as MESH networks are able to connect all points in a network together; A network operations center for very small aperture terminal (VSAT) operation (see also Shared Hub).

- **hybrid satellite**
  A satellite which carries two or more different communications payloads (i.e., C-band and Ku-band).
- **hydrographic and bathymetric surveying**
  Hydrographic and bathymetric surveying is the surveying or mapping of harbors, inlets and deep water locations. Hydrography is the study of the physical characteristics of oceans, lakes and rivers as well as the elements affecting safe navigation. Bathymetry is the measurement and study of water depths.
- **hyperscroll™**
  New HyperScroll™ feature locates and displays fish targets at higher boat speeds with increased soundings per second.
- **icann (internet corporation for assigned names & numbers)**
  organization established by the United States Department of Commerce in 1998 to oversee how Internet names are awarded.
- **icon**
  A graphic symbol that can be placed on the plotter screen (and

saved in memory) to represent some special event or area of interest to the GPS user. Also known as Event Marker.

- **ieee (institute of electrical and electronic engineers, inc.)**

A USA based international technical/professional society founded in 1884. The IEEE promotes standardization in technical disciplines which include aerospace, biomedical technology, computers and communications, consumer electronics and electrical power. The institute conducts educational programs, sponsors international technical conferences and symposia and publishes nearly 25% of the world's technical papers in electronics and computer engineering and computer science.

- **ieee 802**

A set of IEEE standards applicable to local area networks, The equivalent ISO standard is IS 8802:

1. IEEE 802.2: Defines Logical Link Control (LLC, the upper portion of the data link layer) for local area networks
2. IEEE 802.3: Defines the transport layer of (a variant of) Ethernet.
3. IEEE 802.3u: The IEEE committee working on standards for Fast Ethernet. IEEE 802.3z: The IEEE committee working on standards for Gigabit Ethernet. IEEE 802.4: Token Bus standard.
4. IEEE 802.5: Token ring standard.

- **igeb-interagency gps executive board**

IGEB established by Presidential directive in 1996 to manage the Global Positioning System (GPS) and its U.S. Government augmentations as a national asset.

- **illumination**

The signal coming from the GPS Satellites.

- **inclination**

One of the orbital parameters that describes the orientation of an orbit. It is the angle between the orbital plane and a reference plane, the plane of the celestial equator for geocentric orbits and the ecliptic for heliocentric orbits; The angle between the orbital plane of a satellite and the equatorial plane of the earth.

- **inclined orbit**

A condition in which a satellite is unable to maintain a geostationary position above the earth's equator. Almost all satellites generate electrical power to operate their transponders by converting the sun's energy to electricity. Energy used for station keeping, that is keeping the satellite within a very narrow range of movement north or south of the equator, is stored on board the spacecraft and over time is dissipated, usually after passage of the design life of between 8-15 years. When this fuel runs out, the satellite is no longer able to keep station above the equator. Since satellite uplink and downlink antenna are aimed at a particular point above the equator, a satellite in inclined orbit will move in and out of the antenna's "range of vision" as the satellite's orbit fluctuates north and south of the equator.

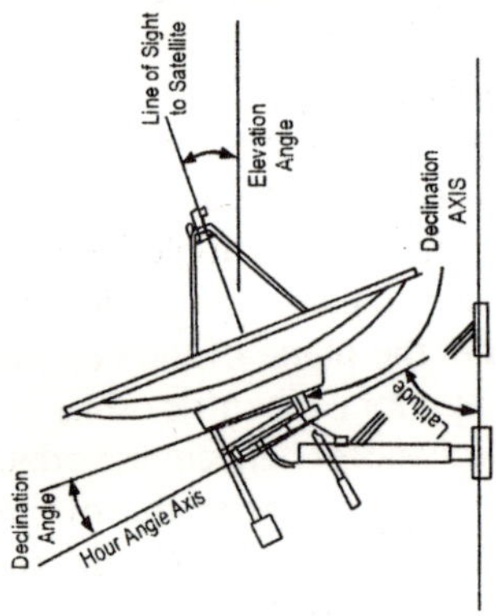

- **increment**

  The change in longitude of the ascending node between two successive passes of a specified satellite, measured in degrees West per orbit.

- **independent baseline**

  These are baselines observed using GPS Relative Positioning techniques which are the minimum necessary to transfer the Datum from one Base Station to all other stations within a ground network. For example, if there are M stations, there will be M-1 independent baselines linking all the stations. Any extra baselines that are measured are "redundant" baselines which may improve the quality and reliability of the station coordinates after Network Adjustment; Non-trivial baseline. Those vectors determined from differencing common phase measurements only once. For any given session there are n-1 independent vectors where n is the number of receivers operating.

- **index contour**

  A contour line on a topographic map that is labeled as to elevation; used as a reference.

- **index of facilities gps**

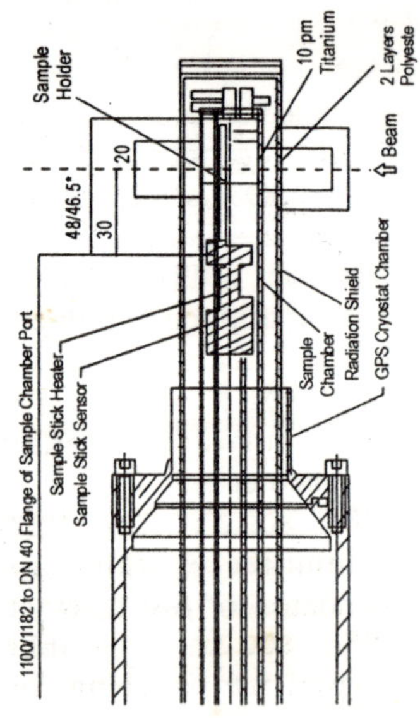

- **initialization**

The first time a GPS receiver orients itself to its current location and collects almanac data. After initialization has occurred, the receiver remembers its location and acquires a position more quickly because it knows which satellites to look for; Entering such data as time, time off-set, approximate position, and antenna height into a receiver (cold start with no almanac) to help the unit find and track satellites; Refers to the procedure of telling a new GPS receiver where it is, when it is turned on for the first time. Information required for initialization includes approximate present position in latitude/longitude coordinates, the current local time and date; The initial orienting process that a G.P.S. receiver does when it is first turned on. If it is the first time the receiver has been turned on or if the receiver is more than a few hundred miles from the last location it was used, it will take some time or help from the operator by inputting or selecting from a menu the approximate location.

- **inmarsat**

An international consortium chartered in the mid-1970s to provide improved maritime public correspondence and radio determination capabilities; The International Maritime Satellite Organization operates a network of satellites for international transmissions for all types of international mobile services including maritime, aeronautical, and land mobile.

- **inmarsat-international mobile satellite organization**

Established in 1979 to serve the maritime industry by developing satellite communications for ship management and distress and safety applications.

- **input/output (i/o)**

The two-way transfer of GPS information with another device, such as a nav plotter, autopilot, or another GPS unit.

- **instructional television fixed service (itfs)**

A service provided by one or more fixed microwave stations operated by an educational organization and used to transmit instructional information to fixed locations.

- **integer bias terms**

The receiver counts the radiowaves from the satellite, as they pass the antenna, to a high degree of accuracy. However, it has no information on the number of waves to the satellite at the time it started counting. This unknown number of wavelengths between the satellite and the antenna is in integer bias term.

- **integrated doppler**

A measurement of Doppler shift frequency of phase over time.

- **integrated services digital network**

A digital telephone service that allows for a combination of voice and data connections over a single high-speed connection.

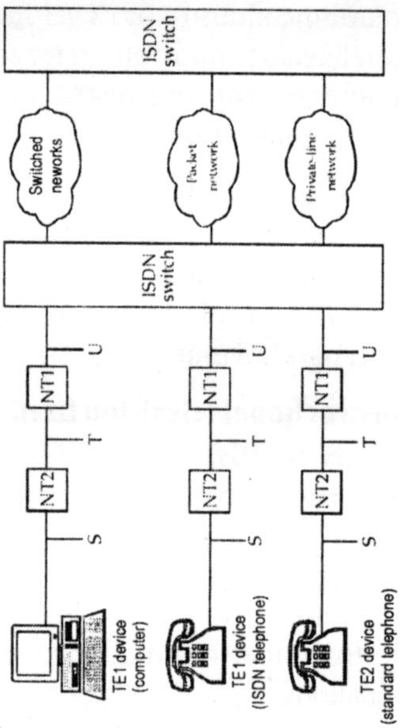

A network that provides end-to-end digital connectivity to support a wide range of services, including voice and non-voice services, to which users have access by a limited set of standard multi-purpose user network interfaces, as defined in the ITU-T1 series.

- **integrity**

A quality measure of GPS performance for critical applications such as civilian aviation. A high level of integrity is sought for such applications; The ability of a system to supply timely warnings in the event of a loss of navigation solution, excessive noise, or other factors affecting measured position.

- **intelligent transportation systems (its)**

Advanced communication technology that enhances the existing highway infrastructure of the facility which improves safety on the highways, increases capacity on the highways, and makes them function at a more efficient level.

- **intelsat**

The International Telecommunications Satellite Organization operates a network of satellites for international transmissions.

- **interactive video data service (ivds)**

A communication system, operating over a short distance, that allows nearly instantaneous two-way responses by using a hand-held device at a fixed location. Viewer participation in game shows, distance learning and e-mail on computer networks are examples.

- **interference**

Any distortion of the transmitted signal that impedes the reception of the signal at the receiver;

Energy which tends to interfere with the reception of the desired signals, such as fading from airline flights, RF interference from adjacent channels, or ghosting from reflecting objects such as mountains and buildings.

- **interferometric differencing**
  A variation of phase differencing in which two different antennas sample the GPS signal wave front at two locations and then feed the information into a single amplifier and interferometer (mixer) in which phase difference observations are made.

- **interlaced scanning**
  The television picture in the NTSC system is made up of 525 scan lines. One-half of the scan lines which make up the video picture are traced on the screen approximately 30 times a second alternating with the other one-half of the scan lines which are traced on the screen approximately 30 times a second thus providing 60 interlaced line scanning events each second.

- **intermodality**
  The ability to move between two or more modes of transport.

- **international gps service (igs)**
  An initiative of the International Association of Geodesy, as well as several other scientific organisations, that was established as a service at the beginning of 1994. The IGS comprises of many component civilian agencies working cooperatively to operate a permanent global GPS tracking network, to analyse the recorded data and to disseminate the results to users via the Internet. The range of "products" of the IGS include precise post-mission GPS satellite ephemerides, tracking station coordinates, earth orientation parameters, satellite clock corrections, tropospheric and ionospheric models. Although these were originally intended for the geodetic community as an aid to carrying out precise surveys for monitoring crustal motion, the range of users has since expanded dramatically, and the utility of the IGS is such that it is vital to the definition and maintenance of the International Terrestrial Reference System.

- **international terrestrial reference system (itrs)**
  The most precise, geocentric, globally-defined coordinate system or datum on the earth's surface. It is a more accurate and more convenient a Satellite-Based Datum than the WGS84 Datum. The various "frames" are realisations of the ITRS for a particular epoch in time, consisting of a set of 3-D coordinates and velocities for hundreds of geodetic stations around the world (all coordinates

of fixed stations on the earth change with time due to "continental drift"). Although some of the stations are Satellite Laser Ranging (SLR) stations, or Very Long Baseline Interferometry (VLBI) stations, the vast majority are GPS tracking stations of the IGS network.

- **internet**

The Internet is the largest interconnection of networks in the world. It is a three level hierarchy composed of backbone networks such as ARPAnet, NSFnet and MILNET, mid-level networks, and stub networks. These include among others commercial (.com or .co), university and educational (.edu or .ac), other research networks (.org or .net), governmental (.gov) and military (.mil) networks and span may different networks around the world with various protocols including the Internet Protocol.

- **internetworking**

The process of interconnecting a number of individual networks to provide a path from a terminal or a host on one network to a terminal or a host on another network. The networks involved may be of the same type, or they may be of different types. However, each network is distinct, with its own addresses, internal protocols, access methods, and administration.

- **invert route**

To display and navigate a route from end to beginning for purposes of returning to the route's starting point.

- **ion**

An electrically charged particle formed by either adding or taking away electrons from neutral particles.

- **ionization**

The process by which atoms form electrically charged particles called ions.

- **ionosphere**

A region of the earth's atmosphere where ionization caused by incoming solar radiation affects the transmission of GPS radio waves. It extends from a height of 50 kilometers (30 miles) to 400 kilometers (250 miles) above the surface; The band of charged particles 80 to 120 miles above the Earth's surface; That layer of the atmosphere approximately 30-300 miles above the earth's surface that contains electrically charged particles (ions). These charged particles interfere with or distort transmissions of electromagnetic signals through the layer; The layers of ionized air in the upper atmosphere (above about 70 kilometers). Depending on frequency, the ionosphere refracts radio signals to cause blockage, alter their path or impact the propagation speed. Navigation signals penetrating the ionosphere are delayed. These delays can be either predicted using (not terribly accurate) models, or measured using dual-frequency receivers.

- **ionosphere, ionospheric delay**

The Ionosphere is that band of atmosphere extending from about 50 to 1000 kilometres above the earth's surface in which the sun's ultraviolet radiation ionises gas molecules which then lose an electron. These free electrons influence the propagation of microwave signals (speed, direction and polarisation) as they pass through the layer. The Ionospheric Delay on GPS signals is frequency-dependent and hence impacts on the L1 and L2 signals by a different amount (unlike that within the Troposphere). A linear combination of pseudo-range or carrier phase observations on the L1 and L2 carrier waves can be created to almost entirely eliminate the Ionospheric Delay. The resulting observable is known as the Ionosphere-Free carrier phase (or pseudo-range). For single-frequency receivers it is not possible to account for this signal bias in this way. A broadcast model is contained within the transmitted Navigation Message, however, it is a relatively poor model (unlikely to account for more than 50% of the effect) as the Delay is very difficult to predict. The magnitude of the Ionospheric Delay is a function of the latitude of the receiver, the season, the time of day, and the level of solar activity. The Delay in the Zenith direction can be several tens of metres, increasing as the elevation angle of the satellite signal reduces (being 3-5 times greater than in the Zenith direction). The Delay is largely eliminated in Relative or Differential Positioning, however, the residual Ionospheric Delay increases as the baseline length increases and may be a significant source of error (especially in the height component) for very high precision GPS Geodesy. Even when using dual-frequency instrumentation, the Ionospheric Delay can still cause problems during the process of rapid Ambiguity Resolution when phase and range combinations other than the Ionosphere-Free one are used.

- **ionospheric refraction**

The change in the propagation speed of a signal as it passes through the ionosphere.

- **ird**

An integrated receiver and decoder for reception of a transmission of voice, video and data.

- **ird (integrated receiver/decoder)**
  Satellite ground segment equipment which receives and decodes digital information transmitted via satellite RF signals. Manufactures have until now incorporated proprietary algorithms into digital transmission systems which result in the lack of interoperability among the differing systems. An IRD will only operate successfully within the particular proprietary system of which it is a part.

- **isa (intelligent speed adaptation)**
  Systems designed to alert drivers when they exceed the speed limit or are travelling dangerously slow, with some systems also offering dynamic correction capabilities.

- **isdn - integrated services digital network.**
  A CCITT standard for integrated transmission of voice, video and data. Bandwidths include: Basic Rate Interface - BR (144 Kbps - 2 B & 1 D channel) and Primary Rate - PRI (1.544 and 2.048 Mbps); A standard for the integrated transmission of voice, video and data developed by the Consultative Committee on International Telephony and Telegraphy (referred to as the CCITT now combined with the CCIR, Consultative Committee on International Radio to form the TSS, Telecommunications Standardization Sector). The ISDN bandwidths include the Basic Rate Interface or BRI at 144 Kb/s and the Primary Rate Interface or PRI at 1.544 and 2.048 MB/s.

- **isl (inter-satellite link)**
  A satellite architecture whereby two or more satellites are configured in such a way that they may communicate directly with one another. This architecture is used in the TDRESS system of satellites and is a component of the design of Big LEO communications satellite systems.

- **iso (international standards organization)**
  An international organization which developed the MPEG and JPEG standards and which is closely allied with the CCITT (part of the Telecommunications Standardization Sector, TSS)

- **isobath**
  A line on a map connecting two points of equal water depth.

- **isogloss**
  A line on a map between linguistically varied areas.

- **isogone**
  A line on a map used to show characteristics of the. earth's magnetic field.

- **isogonic lines**
  A line on a map or chart where the magnetic deviation is the same.

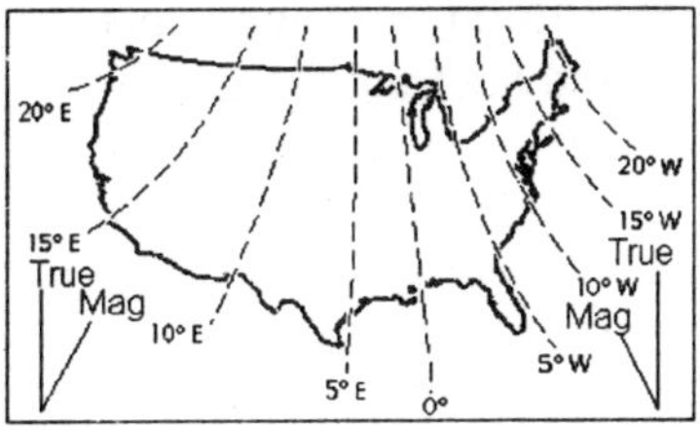

- **isogram**
  A line on a map connecting points of equal value.

- **isograph**
  A line on a map indicating areas that are linguistically similar.

- **isogriv**
  A line drawn on a map such that all points have equal grid variation.

- **isohyet**
  A line on a map connecting points having equal rainfall.

- **isotach**
  A line on a map connecting points of equal wind velocity.

- **isotherm**
  A line on a map connecting points of equal mean temperature.

- **isotropic antenna**
  A hypothetical omni directional point-source antenna that serves as an engineering reference for the measurement of antenna gain.

- **its (intelligent transport systems and services)**
  Any system or service that makes the movement of people or goods more efficient and economical, thus more "intelligent".

- **itu (international telecommunications union)**
  A United Nations treaty organization(originally created in 1865 as the International Telegraph Union with a membership of twenty nations...current name adopted in 1932) which supports procedures for the international allocation of the radio frequency (RF) spectrum and provides the platform for the World Radio Conference (WRC), a biannual meeting of world communication leaders. The WRC publishes 'International Radio Regulations' for the RF spectrum. The ITU conducts ongoing policy and study group sessions.

- **itu-t multimedia teleconferencing standards**
  Standards developed and ratified by the Telecommunication Standardization Sector of the International Telecommunications Union which define the core technologies for multimedia teleconferencing. Standards are developed to promote interoperability among different proprietary systems. These core standards include: T.120: Data protocols for Real Time Data Conferencing (Audiographics). These standards cover document conferencing and application

sharing. H.320: Governs the basic video-telephony concepts of audio, video and graphical communications over circuit-switched media such as ISDN. H.323: An extension of the H.320 Standard to provide for multimedia telecommunications over intranets and packet-swithched networks generally -- applicable to the use of video and other communications over the Internet. Provides standards for both reliable and unreliable transmission over the Internet. H.324: Addresses and specifies a common method of sharing video, data, and voice simultaneously using V.34 modem connections over single analog POTS (Plain Old Telephone Service) line.

- **ixc (interexchange carrier)**

Provides long-distance transport services between points-of-presence (POP) that are established for each local exchange carrier.

- **jammer**

An active electronic counter-measures (ECM) device designed to deny intelligence to unfriendly detectors or to disrupt communications.

- **jpals**

Joint Precision Approach and Landing System, is a DGPS system pro rapid deployment. Uses WAAS/LAAS.

- **jpals-joint precision approach and landing system**

A differential GPS that will provide a rapidly deployable mobile, adverse weather, adverse terrain, day-night, and survivable precision approach and landing capability that supports joint service, civil, and multi-national interoperability.

- **jpeg**

ISO Joint Picture Expert Group standard for the compression of still pictures; A subgroup of the ISO which has established international standards for the digital compression of still pictures.

- **jpo**

Joint Program Office for GPS located at the United States Air Force Space Division in California. Contains members of a number of government agencies with an involvement in controlling the GPS system; Joint Program Office for GPS located at the USAF Space Division at El Segundo, California. The JPO consists of the USAF Program Manager and Deputy Program Managers representing the Army, Navy, Marine Corps, Coast Guard, Defense Mapping Agency and NATO.

- **julian date**

The number of days that have elapsed since 1 January 4713 B.C. in the Julian calendar. GPS time zero is defined to be midnight,

Universal Time Coordinated (UTC), Saturday/Sunday, 6 January 1980 at the 0-degree meridian at Greenwich, London, England. The Julian date for GPS time zero is 2,444,244.5.

- **jupiter gps module**

minmal rs232 interface for laptop pc (tested with Dell laptop)

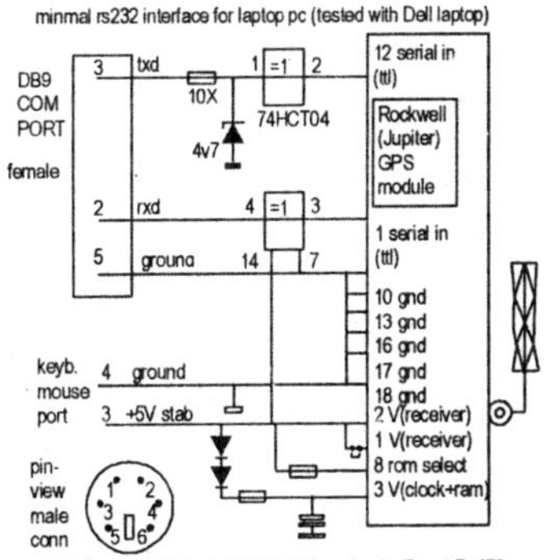

Batt si a 3.6V NiCd cell, 2* 1N4148 in series for D and R=470 ohm. If you use a supercap put it on the C2 position of the jupiter board and connect pin 3 directly to+5V.

This is a terminal circuit with no safety precaution whatsover!
Connect pin1 only when using an active 5V antenna
This circuit only works with TTL (0V/5V)compatible serial ports

- **ka band**

A portion of the RF spectrum located between 18 GHz and 31 GHz. Downlink frequencies for satellite communications are located in the 20 GHz range and uplink frequencies are located in the 30 GHz range; A higher frequency than Ku-band, operating from 18 to 31 GHz.

- **kalman filter**

A numerical method used to track a time-varying signal in the presence of noise. If the signal can be characterized by some number of parameters that vary slowly with time, then Kalman filtering can be used to tell how incoming raw measurements should be processed to best estimate those parameters as a function of time; An optimum mathematical procedure for recursively estimating dynamically changing parameters, such as the position and velocity of a vessel, from noise-contaminated observations.

- **karst**

An area of irregular limestone in which erosion has produced fissures, sinkholes, underground streams and caverns.

- **kbps**

Kilobits per second. Refers to transmission speed of 1,000 bits per second.

- **kelvin (k)**

The temperature measurement scale used in the scientific community. Zero K represents absolute zero, and corresponds to minus 459 degrees Fahrenheit or minus 273 Celsius. Thermal noise characteristics of LNA are measured in Kelvins.

- **keplerian elements**

A set of six parameters that describe the position and velocity of a satellite in a purely elliptical (Keplerian) orbit. These parameters are the semimajor axis and eccentricity of the ellipse, the inclination of the orbit plane to the celestial equator, the right ascension of the ascending node of the orbit, the argument of perigee, and the time the satellite passes through the perigee; The classical set of six orbital element numbers used to define and compute satellite orbital motions. The set is comprised of inclination, Right Ascension of the Ascending Node (RAAN), eccentricity, argument of perigee, mean anomaly, and mean motion, all specified at a particular epoch or reference year, day, and time. Additionally, a decay rate or drag factor is usually included to refine the computation.

- **kilohertz**

  A radio signal that has 1,000 cycles per second.

- **kinematic**

  Pertaining to motion or moving objects.

- **kinematic gps**

  Observations while a receiver is in motion. In surveying applications, kinematic refers to uninterrupted carrier-phase measurements following successful solution of the integer ambiguities. This can be accomplished in a continuous mode where the receiver remains in motion for precise positioning of a vehicle, or in an intermittent mode where data is recorded only after a receiver is brought to a stationary point, and the observations while in motion are tracked as a way to maintain the integer ambiguities.

- **kinematic positioning**

  Kinematic positioning refers to applications in which the position of a non-stationary object (automobile, ship, bicycle) is determined; Positioning a continuously moving platform by using GPS carrier-phase data while operating in a differential mode.

- **kinematic surveying**

  Method of surveying that involves starting from a known baseline and tracking a minimum of four satellites. One receiver remains static at a control point while other receivers are moved around to obtain measurements; A form of continuous differential carrier-phase surveying requiring only short periods of data observations.

  Operational constraints include starting from or determining a known baseline, and tracking a minimum of four satellites. One receiver is statically located at a control point, while others are moved between points to be measured; A precision differential GPS surveying technique in which the roving user does not need to stop to collect precision information. Meter to centimeter-level accuracy is available using mode, dual-frequency, carrier-phase measurement techniques; Kinematic surveying is a method that initially solves wavelength ambiguities and retains the resulting measurements by maintaining a lock on a specific number of satellites throughout an entire surveying period.

- **klystron**

A microwave tube which uses the interaction between an electron beam and the RF energy on microwave cavities to provide signal amplification. The klystron operates on principles of velocity modulation very similar to those in a TWT except that klystron interaction takes place at discrete locations along the electron beam. Common types of klystrons are the reflex klystron (an oscillator having only one cavity), two-cavity klystron amplifiers and oscillators, and multi-cavity klystron amplifiers.

- **knot**

A unit of speed of one nautical mile per hour, equal to 1.15 statute mile per hour (1.85 kilometers per hour).

- **ku band**

A portion of the RF spectrum located between 10.9 GHz and 17 GHz, a part of which is dedicated to satellite communications. Satellite downlink frequencies are located between 11.7 GHz and 12.2 GHz and uplink frequencies are located between 14 GHz and 14.5 GHz.

- **l band**

The radio frequencies that extend from 390 MHz to 1550 MHz. The GPS carrier frequencies are in the L band (1227.6 MHz and 1575.42 MHz); Range of radio frequencies covering the band from 390MHz to 1550MHz; The group of radio frequencies extending from 390 MHz to 1550 MHz. The GPS carrier frequencies L1 (15735 MHz) and L2 (1227.6 MHz) are in the L-band. The group of radio frequencies extending from 390 MHz to 1550 MHz. The GPS carrier frequencies L1 (15735 MHz) and L2 (1227.6 MHz) are in the L-band; The group of radio frequencies extending from 390MHz to 1550MHz. The GPS carrier frequencies L1 and L2 are in the L-Band; The frequency range from 0.5 to 1.5 GHz. Also used to refer to the 950 to 1450MHz used for mobile communications; A portion of the RF spectrum located between 500 MHz and 1500 MHz. The RF (radio frequency) spectrum between 950 MHz and 1450 MHz is dedicated to mobile communications; The segment of the microwave portion of the radio spectrum nominally between 1 and 2 GHz.

- **l1**

The primary L-band signal radiated by each NAVSTAR satellite at 1575.42 MHz. The L1 beacon is modulated with the C/A and P-codes, and with the NAV message. L2 is centered at 1227.60 MHz and is modulated with the P-code and the NAV message.

- **l1 & l2**
  L1 and L2 are designations of the two basic carrier frequencies transmitted by GPS satellites that contain the navigation signals. The L1 frequency is 1,575.42 Mhz, while the L2 frequency is 1,227.60 Mhz; These designate the two basic carrier frequencies transmitted by GPS satellites that contain the navigation signals. L1 is 1,575.42 MHz; L2 is 1,227.60 MHz.

- **l1 frequency**
  1575.42 MHz GPS carrier frequency which contains only encrypted P code, used primarily to calculate signal delays caused by the ionosphere; One of the two radio frequencies transmitted by the GPS satellites. This frequency carries the Coarse Acquisition Code (C/A code), P-Code, and the nav message, and is transmitted on a frequency of 1575.42 MHz.

- **l1 frequency band**
  One of the two ultra high radio frequencies transmitted by NAVSTAR GPS satellites. This carrier frequency of 1575.42 MHz transmits both C/A (Standard Positioning Service) and P (Precise Positioning Service) code.

- **l1 signal / frequency band**
  The primary L-band signal transmitted by each GPS satellite at 1572.42 MHz. The L1 broadcast is modulated with the C/A and P-codes and with the navigation message.

- **l2 frequency**
  One of the two radio frequencies transmitted by the GPS satellites. This frequency carries only the P-Code, and is transmitted on a frequency of 1227.6 MHz; 1227.60 MHz GPS carrier frequency which contains only the encrypted P-Code (or Y-Code) and the Navigation Message. Military Y-Code capable receivers can, in addition to making L1 measurements, make pseudo-range measurements on the L2 carrier. The combination of the two measurements (on L1 and L2) permits the Ionospheric Delay to be corrected for. Dual-frequency GPS receivers intended for Surveying applications can make L2 measurements using proprietary signal processing techniques. Such measurements are essential if the Ionospheric Delay on carrier phase is to be corrected for (especially on baselines of length greater than about 20-30km) and/or where fast Ambiguity Resolution is needed.

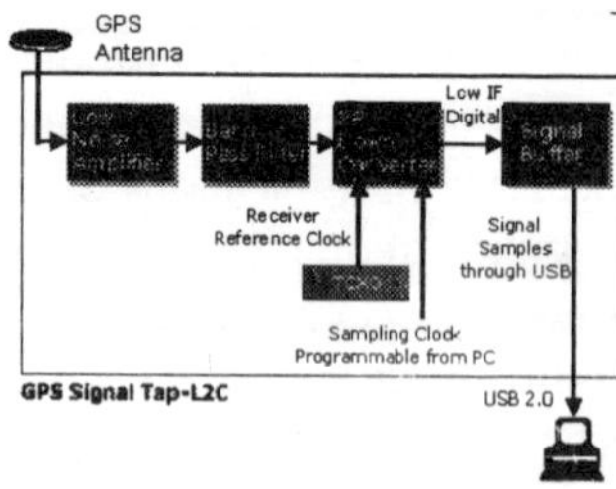

- **l2 signal / frequency band**

The second L-band signal is centered at 1227.60 MHz and carries the P-code and navigation message.

- **laas**

Local Area Augmentation System. A system similar to WAAS, in that similar correction data are used. But in this case, the correction data are transmitted from a local source, typically at an airport or another location where accurate positioning is needed. These correction data are typically useful for only about a thirty to fifty kilometer radius around the transmitter.

- **lan (local area network)**

A bus or ring connected, limited distance network that serves the data communication needs of users within a building or several buildings in proximity to each other. Ethernet is an example of a standard Local Area Network.

- **land mobile service**

A public or private radio service providing two-way communication, paging and radio signaling on land.

- **landline**

Traditional wired phone service.

- **landmark**

A specific location or identifiable natural or man-made geographic feature. Also referred to as a waypoint in G.P.S. navigation.

- **lane**

The area (or volume) enclosed by adjacent lines (or surfaces) of zero phase of either the carrier beat phase signal, or of the difference between two carrier beat phase signals. On the earth's surface a line of zero phase is the locus of all points for which the observed value would have an exact integer value for the complete instantaneous phase measurement. In three dimensions, this locus becomes a surface.

- **lane control signal (lcs)**

Distinctive rectangular signals mounted above each lane displaying symbols intended to guide motorists into the appropriate lanes on the freeways.

- **lat/lon**

Coordinate system using latitude and longitude coordinates to define a position on the earth.

- **lata (local access and transport area)**

A grouping of exchange areas served by a single Local Exchange Carrier (LEC). Traffic between LATAs must be carried by an Inter-Exchange Carrier (IXC).

- **latitude**

A north/south measurement of position perpendicular to the earth's polar axis; A position's distance north or south of the equator, measured by degrees from zero to 90. One minute of latitude equals one nautical mile; A north/south angular measurement of position relative to the equator, in the meridian plane which contains the earth's rotation axis; An angular measurement (distance) of a point on the earth, north or south of the equator. The distance is measured in degrees, minutes, and seconds. Latitude is 0 degrees at the equator, +90 degrees at the North Pole, and -90 degrees at the South Pole (1 degree of latitude equaling 60 nautical miles and 1 minute of latitude being 1 nautical mile. Latitude is constant on a parallel). Lines (parallels) of latitude circle the earth horizontally and are parallel to one another.

- **l-bands or l1/l2 bands**

  A group of radio frequencies that extend from 390 MHz to 1550 MHz. The GPS carrier frequencies (1227.6 MHz and 1575.42 MHz) are also in the L band.

- **lbs (location-based services)**

  Services that depend on knowing the user's location to deliver user-specific information, such as real-time traffic updates or point-of-interest data; A service which uses knowledge provided about where a user is located in order to deliver information relevant to the user's position.

- **lcd (liquid-crystal display)**

  An electronic display panel that is based on the properties of a liquid-crystal material sandwiched between two transparent plates and illuminated from behind. The transmittance of the liquid-crystal material changes with the applied electric field.

- **ldgps**

  Local Differential GPS. Two or more GPS Receivers are used to create a local reference to each other.

- **leased line**

  A dedicated circuit typically supplied by the telephone company.

- **lec (local exchange carrier)**

  Provides public, switched telephone services within limited areas called Local Access and Transport Areas (LATA). The LEC provides connections between users located within the same exchange area, between users located in different exchange areas that are in the same LATA, and delivers traffic directed to an exchange area outside the LATA to Inter- Exchange Carriers (IXC).

- **leg**

  The portion of a route between waypoints.

- **leg (route)**

  A portion of a route consisting of a starting (from) waypoint and a destination (to) waypoint. A route that is comprised of waypoints A, B, C, and D would contain three legs. The route legs would be from A to B, from B to C, and from C to D; A segment of a route that has a starting (FROM) waypoint and a destination (TO) waypoint. A route may consist of 1 or more legs. A route that is from waypoint A to waypoint B to waypoint C to waypoint D has three legs with the first being from waypoint A to waypoint B.

- **legend**

Information on a map for such things as scale, datum, and other details.

- **leo (low earth orbit)**

Until recently, a distinction was made between LEO and MEO (Medium Earth Orbit) orbital classifications. LEO was classified as an earth orbit with an altitude of between 200-2400 kms. The MEO designation classified orbits between 2,400 and 10,000 km. Current literature frequently refers to any orbital distances from earth of less than that for geostationary orbit as being LEO. Almost all of the Big LEO and Little LEO systems which are currently planned will operate in orbits ranging between 640 km and 2,500 km above the earth. Satellites which are placed in LEO orbits move rather swiftly in relation to the earth, generally from a westerly to an easterly direction. Some of the newly designed LEO communications satellite networks will launch some of the satellites in the network constellation into polar earth orbits. The relative motion of the satellite in relation to the earth slows as the altitude of the satellite increases. The rapid relative movement in low earth orbit enables data gathering and communication satellites to cover large areas of the earth's surface in short periods of time. The space shuttle is injected into a low earth orbit, for example:

BIG LEO: A space segment architecture which consists of a constellation of many satellites in Low Earth Orbit in a configuration which will permit the delivery of global mobile telephony and data services. The system is served by a network of ground stations which provides gateway access from terrestrial networks as well as management, control and orbital correction functions. This system is intended to provide telecommunication service to remote sites, high latitude geographic locations which are not accessible by GEO satellites and to omni directional antenna on handset transceivers. Examples of Big LEO systems include Teledesic and Iridium.

LITTLE LEO: A space segment architecture which is similar to that of Big LEO systems but which carries data only. The remote collection and transmission of utility meter data would be an example of a service provided by a Little LEO satellite system. Often, store and forward technology is used to facilitate communication with the ground segment.

- **lidar-also called laser radar**
  A method of detecting distant objects and determining their position, velocity, or other characteristics by analysis of pulsed laser light reflected from their surfaces.

- **line of position (lop)**
  Locus of Points have a constant measurement (such as range, range difference). A fix is determined by crossing two lines of position.

- **line of sight (los) propagation**
  Of an electromagnetic wave, propagation in which the direct transmission path from the transmitter to the receiver is unobstructed. The need for LOS propagation is most critical at GPS frequencies.

- **liquid crystal display (lcd)**
  A display circuit characterized by a liquid crystal element sandwiched between two glass panels. Characters are produced by applying an electric field to liquid crystal molecules and arranging them to act as light filters.

- **lithium battery**
  A soft, silvery, highly reactive metallic element that is used in batteries where weight and cold weather conditions are concerns.

- **lmds (local multipoint distribution system)**
  A fixed, broadband point-to-multipoint wireless service which can provide one-way and two-way high capacity voice, video and data service. Services are licensed and bandwidth auctioned by the FCC.

- **lna**
  Low Noise Amplifier

- **local area augmentation system**
  A specialized type of DGPS to support aircraft landings in a local area (32KM); The implementation of ground-based DGPS to support aircraft landings in a local area (20-mile range); Plan by which Local Area Differe al GPS (LADGPS), which generates and transmits differential corrections to appropriately equipped aircraft users, is augmented with integrity messages transmitted from the ground and additional ranging signals. LAAS is set up near a major airport, and consists of the DGPS reference station, the integrity monitoring receiver and

a pseudolite which transmits "satellite-like" PRN-coded signals to incoming aircraft.

- **local area dgps (ladgps)**
  A form of DGPS in which the user? GPS receiver receives real-time pseudorange and, possibly, carrier- phase corrections from a reference receiver generally located within line of sight. The corrections account for the combined effects of navigation message ephemeris and satellite clock errors (including the effects of SA), and usually, atmospheric propagation delay errors at the reference station.

- **local area network (lan)**

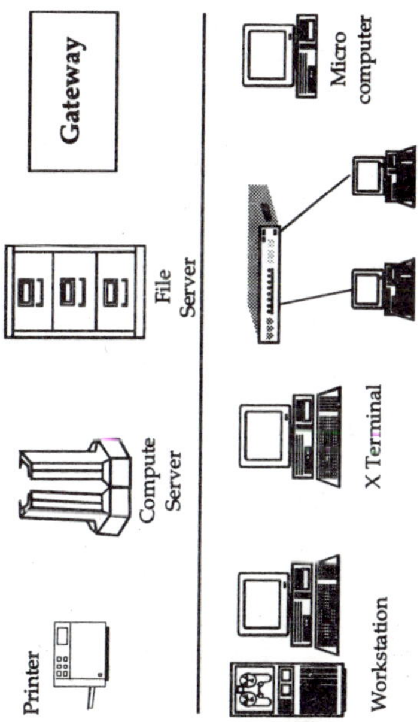

A data communications system that:

(a) lies within a limited spatial area,
(b) has a specific user group,
(c) has a specific topology, and
(d) is not a public switched telecommunications network, but may be connected to one.

- **longitude**
  An east/west measurement of position in relation to the Prime Meridian, an imaginary circle that passes through the north and south poles; The distance east or west of the prime meridian (measured in degrees). The prime meridian runs from the north to south pole through Greenwich, England; The angular measurement of a point on the earth's surface, east or west of the prime meridian. The prime meridian runs through Greenwich, England and is 0 degrees longitude. Since measurements are made East and West, the maximum longitude value is 180 degrees. Mathematically, longitudes are usually denoted as positive for easterly longitudes (e.g., +71 degrees = 71 E), and negative for westerly longitudes (e.g., -65 degrees = 65 W).

- **loran**
  Loran, which stands for Long Range Navigation, is a grid of radio waves in many areas of the globe that allows accurate position plotting. Loran transmitting

stations around the globe continually transmit 100 kHz radio signals. Special shipboard Loran receivers interpret these signals and provide readings that correspond to a grid overprinted on nautical charts. By comparing signals from two different stations, the mariner uses the grid to determine the position of the vessel; Long range navigation system that determines position by comparing the arrival times of radio signals with two or more master/secondary station pairs.

- **loran-c**
  Long range navigation system that determines position by comparing the arrival times of radio signals with two or more master/secondary station pairs.

- **low noise amplifier (lna)**
  This is the preamplifier between the antenna and the earth station receiver. For maximum effectiveness, it must be located as near the antenna as possible, and is usually attached directly to the antenna receive port. The LNA is especially designed to contribute the least amount of thermal noise to the received signal.

- **low noise block down converter (lnb)**
  A combination Low Noise Amplifier and down converter built into one device attached to the feed.

- **low power fm radio (lpfm)**
  A broadcast service that permits the licensing of 50-100 watt FM radio stations within a service radius of up to 3.5 miles and 1-10 watt FM radio stations within a service radius of 1 to 2 miles.

- **low power television (lptv)**
  A broadcast service that permits program origination, subscription service or both via low powered television translators. LPTV service includes the existing translator service and operates on a secondary basis to regular television stations. Transmitter output is limited to 1,000 watts for normal VHF stations and 100 watts when a VHF operation is on an allocated channel.

- **low-power satellite**
  A satellite with less than 30 Watts of transponder radio frequency (RF) transmitting power.

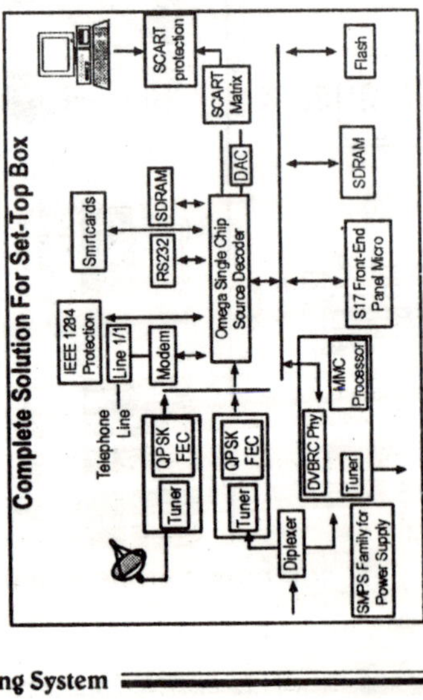

- **luminance**

  In color video systems, a component signal that represents the brightness of the image.

- **mac (a, b, c, d2)**

  Multiplexed analog component color video transmission system. Subtypes refer to the various methods used to transmit audio and data signals.

- **mag. var.**

  Magnetic Variation the different between true North (pointing towards the Geographic Pole) and Magnetic North (pointing towards Magnetic Pole) where a compass points to. The magnetic variation of the earth changes at a rate of 50.27 seconds of arc per year. Memory trick, East is Least and West is Best (You subtract East variations and Add West variations)

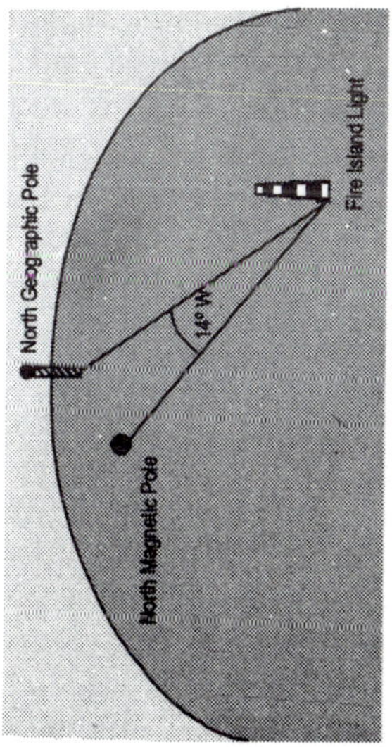

Magnetic Variation

- **magic (mobile automotive geographic information core)**

  An industry consortium established to develop and promote an open industry specification for delivering navigation, telematics and related geographic information services across multiple networks, platforms, and devices.

- **magnetic declination**

  The difference, in degrees, between magnetic north and true north; The difference between True North and the direction that a magnetic compass points for a given location. Magnetic declination varies from place to place and can change as much as a degree in a year.

- **magnetic deviation**

  The error in a magnetic compass reading caused by the difference between the true north pole and the magnetic north pole, also referred to as magnetic variation.

- **magnetic north**

  Represents the direction of the north magnetic pole from the observer's position. The direction a compass points; The direction to the Magnetic North Pole. It is what a magnetic compass indicates. It is different from True North, by the value of the Mag. Var.; Magnetic north is the location our compasses point to; it lies several hundred miles to the south of true north, at a location in Canada.

- **maidenhead grid squares**
  A system designed to facilitate the designation of geographical positions for use within the amateur radio community.

- **map datum**
  What reference map is used in determining the Fixes; Since the earth is not flat a method must be used to create a "flat map" from a round earth. Maps and charts are based on a survey of the area that's covered by the map or chart. These surveys are called "Datums". Maps that are created using different datums will show the same latitude/longitude in slightly different locations. Different datums are based on different mathematical models of the earth's shape and dimensions (ELLIPSOIDS) plus an additional factor of PROJECTION.

- **map display**
  A graphic representation of a geographic area and its features.

- **map projection**
  The systematic arrangement of the earth's spherical or geographic coordinate system onto a plane; the process of transforming a globe into a flat map with the least amount of distortion; a transformation process; The method used to convert the spherical shape of the earth onto the plane of a map; A systematic representation of the (spherical) earth on a flat (plane) surface. each map projection has specific properties that make it useful for specific purposes. projections can show true directions (aximuthal), true distances (equidistant), true areas & size (equal area) or true shapes & angles (conformal). one or more but never all; there is always some distortion.

- **map projection deformation**
  The error or distortion that results when the spherical shape of the earth is projected onto the plane of a map.

- **map scale**
  The scale of a map is the ratio between a distance on the map and the corresponding distance on the earth, with the distance on the map typically expressed as 1. thus, a scale of 1:100,000 means 1 inch on the map equals 100,000 inches (approximately 1.6 miles) on the earth. large scale maps depict a small area and show more detail. small scale maps depict a large area and show less detail. for example, a map of bozeman might show every city street, while a map of montana might show only highways. the locational accuracy of a geographic file depends on the scale of the original source map. for example, suppose the person digitizing a map traces the map features within 1/32 of an inch. 1/32 of an inch represents about 1/2 mile at a scale of 1:1,000,000;

650 feet (.12 mile) at 1:250,000; and 62.5 feet at 1:24,000.

- **margin**
  The amount of signal in dB by which the satellite system exceeds the minimum levels required for operation.

- **mask angle**
  The minimum acceptable satellite elevation above the horizon to avoid blockage of line-of-sight; The point above the observer's horizon below which satellite signals are no longer tracked and/or processed. 10° to 25° is typical.

- **master antenna television (matv)**
  An antenna system that serves a concentration of television sets such as in apartment buildings, hotels or motels.

- **mcx**
  Antenna connector used on some of the newer GPS units.

- **mean anomaly**
  An angle that increases uniformly with time, starting at perigee. Used to indicate where a satellite is located along its orbit. Mean Anomaly is usually specified at the reference epoch time where the Keplerian Elements are defined.

- **mean motion**
  The Keplerian Element that indicates the complete number of orbits a satellite makes in a day.

- **mean sea level**
  The average level of the ocean's surface, as measured by the level halfway between mean high and low tide. Used as a standard in determining land elevation or sea depths; The average height of the sea surface, based upon hourly observation of the tide height on the open coast or in adjacent waters that have free access to the sea. In the United States, it is defined as the average height of the sea surface for all stages of the tide over a nineteen year period. Mean sea level, commonly abbreviated as MSL and referred to simply as 'sea level,' serves as the reference surface for all altitudes in upper atmospheric studies.

- **medium-power satellite**
  A satellite with greater than 30 Watts but less than 100 Watts of transponder radio frequency (RF) transmitting power.

- **megahertz (mhz)**
  One million cycles per second. Used to describe a radio frequency.

- **meridian**
  An imaginary line that circles the earth, passing through the geographic poles and any given point on the earth's surface. All points on a given meridian have the same longitude; An imaginary

line circling the earth, passing through the north and south poles; a line of longitude.

- **mgrs**

Military Grid Reference System. The MGRS is an alphanumeric version of a numerical UTM (Universal Transverse Mercator) or UPS (Universal Polar Stereographic) grid coordinate.

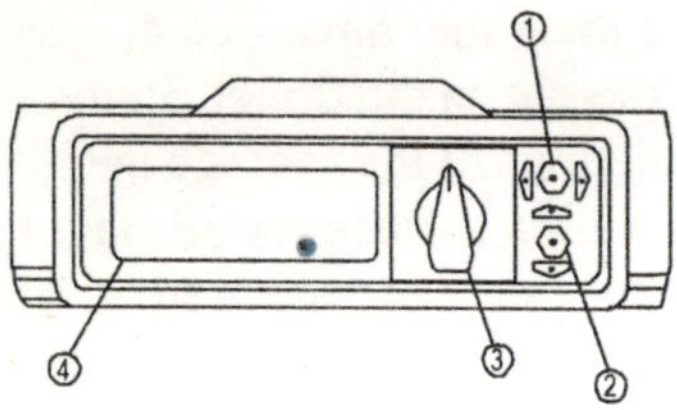

- **microstrip antenna**

A type of antenna commonly used with GPS receivers. It is usually constructed of one or more (typically rectangular) elements that are photoetched on one side of double-coated, printed-circuit board.

- **microwave**

Line-of sight, point-to-point transmission of signals at high frequency. Many CATV systems receive some television signals from a distant antenna location with the antenna and the system connected by microwave relay. Microwaves are also used for data, voice, and indeed all types of information transmission. The growth of fiber optic networks have tended to curtail the growth and use of microwave relays.

- **microwave interference**

Interference which occurs when an earth station aimed at a distant satellite picks up a second, often stronger signal, from a local telephone terrestrial microwave relay transmitter. Microwave interference can also be produced by nearby radar transmitters as well as the sun itself. Relocating the antenna by only several feet will often completely eliminate the microwave interference.

- **mid band**

A portion of the VHF (Very High Frequency) RF spectrum located between television channels 6 and 7 (88 MHZ to 174 MHZ) which has been reserved by the FCC for air, maritime and land mobile units, FM radio and aeronautical and maritime navigation. The frequencies between 108 MHZ and 174 MHZ can be used to provide additional channels on cable television systems.

- **minimally constrained**

A form of least squares solution in which the observed baseline vectors are treated as "observations" in a secondary network adjustment, and only one coordinate must be held fixed to its known value while all others are allowed to adjust. Typically GPS surveys measure more baselines than the minimum needed to coordinate all the points in the network. These extra "observations" are redundant information that a minimally constrained network adjustment uses to derive optimum estimates of the coordinate parameters, as well as valuable quality information in the form of parameter standard deviations and error ellipses (or ellipsoids).

- **minute**

A unit of measurement; one degree contains 60 minutes.

- **mob-man over board**

A feature in G.P.S. receivers to quickly mark a given location as you are traveling. Some receivers then activate a GOTO navigation function to that waypoint.

- **modem**

This term is a contraction of "modulator" and "demodulator". A modem coverts digital data generated by and coming into a computer which are transmitted over telephony circuits or by using terrestrial or satellite RF (radio frequency) circuits; Acronym for Modulator-Demodulator. A device that modulates and demodulates signals. Note: Modems are primarily used for converting digital signals into quasi-analog signals for transmission over analog communication channels and for reconverting the quasi-analog signals into digital signals.

- **modulation**

A method of encoding a message signal on top of a carrier, which can be decoded at a later time; The process of manipulating the frequency or amplitude of a carrier in relation to an incoming video, voice or data signal; A method of encoding a message signal on top of a carrier, which can be decoded at a later time. Modulation: the process of modifying the channel signal (often called the "carrier") to represent information to be transmitted. typical modulation methods involve the frequency, amplitude or phase of the carrier.

- **modulator**

A device which modulates a carrier. Modulators are found as components in broadcasting transmitters and in satellite transponders. Modulators are also used by CATV companies to place a baseband video television signal onto a desired VHF or UHF channel. Home video tape recorders also have built-in modulators which enable the recorded video information to be played back using a television receiver tuned to VHF channel 3 or 4.

- **monitor station**
  Worldwide group of stations used in the GPS control segment to monitor satellite clock and orbital parameters. Data collected here are linked to a Master Station where corrections are calculated and controlled. These data are uploaded to each satellite at least once per day from an Upload Station.

- **mops**
  Minimum Operational Performance Standards.

- **most cooperation**
  Cooperation between car manufacturers and OEM suppliers to develop a low cost, high bandwidth solution for data and multimedia in-vehicle applications.

- **moving map**
  A program that displays your position in real time on the top of either a Vector or Raster map using GPS input.

- **mpeg**
  The Moving Pictures Experts Group, the television industry's informal standards group; A working group of a committee of the ISO (International Standards Organization). They developed the MPEG worldwide standard for motion video compression and transport.

- **mpeg-2**
  The agreed standard covering the compression of data (coding and encoding) for digital television. MPEG-2 changed the meaning behind the aspect_ratio_information variable, while significantly reducing the number of defined aspect ratios in the table. MPEG-2 can represent interlaced or progressive video sequences

- **msas**
  Multi-functional Satellite Augmentation System, operation in Asia.

- **msas-mtsat satellite-based augmentation system**
  Japanese version of WAAS-covers parts of Asia and the Pacific

- **mss**
  Mobile Satellite Service

- **mtops (millions of theoretical operations per second)**
  A measure of computer processing speed used by the US Department of Commerce (USDoC) to define a classification system for export controls applicable to high-speed computers. For example, in February, 2000, the USDoC eliminated export controls on computers with speeds below 12,300 MTOPS to all countries with the exception of Iraq, Libya, North Korea,Cuba, Sudan and Syria.

- **mtsat-multi-functional transport satellite**

  A satellite with two main functions: a meteorological mission conducted by Japan Meteorological Agency; an aeronautical mission conducted by Japan Civil Aviation Bureau of the Ministry of Land, Infrastructure and Transport of Japan (MLIT).

- **multicast**

  Multicast is a subset of broadcast that extends the broadcast concept of one to many by allowing "the sending of one transmission to many users in a defined group, but not necessarily to all users in that group."

- **multichannel receiver**

  A receiver containing many independent channels that offers the best level of performance because each channels tracks one satellite continuously; A GPS receiver that can simultaneously track more than one satellite signal; High quality receivers may have 12 channels for L1, and another 12 channels for L2 signals. Lower quality GPS navigation receivers may have only 6 or 8 channels. In contrast to a Multiplexing Channel Receiver.

- **multipath**

  Interference caused by reflected GPS signals arriving at the receiver, typically as a result of nearby structures or other reflective surfaces; The interference to a signal that has reached the receiver antenna by multiple paths; usually caused by the signal being bounced or reflected. Signals from satellites low on the horizon will have high multipath error. Receivers that can be configured to "mask out" signals from such satellites can help minimize multi-path; Signal arrival at a receiver? antenna by way of two or more different paths such as a direct, line-of-sight path and one that includes reflections off nearby objects. The difference in path lengths causes the signals to interfere at the antenna and can corrupt the receiver? pseudo range and carrier-phase measurements.

- **multipath error**

  Errors caused by the interference of a signal that has reached the receiver antenna by two or more different paths. This is usually caused by one path being bounced or reflected. The impact on a pseudo-range measurement may be up to a few metres. In the case of carrier phase, this is of the order of a few centimetres.

- **multiple access**

  Techniques have been developed in the satellite industry which allow satellite spectrum and power to be shared efficiently among multiple users of a satellite transponder.

CDMA (CODE DIVISION MULTIPLE ACCESS): A multiple access scheme whereby ground station uplinks access a satellite transponder using spread-spectrum modulations and orthogonal codes to avoid interfering with other transmissions using the same transponder. In contrast with the FDMA scheme which attempts to minimize the transmitted bandwidth, in this scheme all users transmit signals simultaneously across all of the dedicated multiple access channel. Receivers use a code corresponding to the transmission code to demodulate the signal or separate it from other signals on the channel.

FDMA (FREQUENCY DIVISION MULTIPLE ACCESS): A multiple access scheme whereby each ground station uplink is assigned a specific frequency slot and bandwidth for one of the multiple carriers within a specific satellite transponder. This scheme is usually used in conjunction with Frequency Modulation. The FDMA scheme may be divided into two categories, Multiple Channel Per Carrier and Single Channel Per Carrier.

TDMA (TIME DIVISION MULTIPLE ACCESS): A multiple access scheme whereby many users may access a single carrier by time sharing. A digital signal is compressed in packets which are transmitted to the carrier in bursts. These packets are processed into consecutive time segments which do not overlap. Instructions are sent to a receiver which identify the packets representing a particular transmission (ie., every 4th packet).

- **multiplexing**

The technique used in some GPS receivers of rapidly sequencing the signals of two or more satellites through a tracking channel. This ensures navigation messages from the satellites tracked by the channel are essentially acquired simultaneously; Multiplexing is a technique used in some global navigation satellite receivers to sequence the signals of two or more satellites through a single hardware channel. Multiplexing allows a receiver to track more satellites than the number of hardware channels at the cost of lower effective signal strength; A process in the transmission of RF (radio frequency) signals whereby a number of simultaneous transmissions can be accommodated on a single circuit; The technique used in some GPS receivers of rapidly sequencing the signals of two or more satellites through a tracking channel. This ensures navigation messages from the satellites tracked by the channel are essentially acquired simultaneously.

- **multiplexing channel**

A channel of a GPS receiver that can be sequenced through a number of satellite signals. In contrast to a Multi-Channel Receiver in which one channel is dedicated to each satellite signal; A channel of a GPS receiver that can be sequenced through a number of satellite signals; A receiver channel which is sequenced through several satellite signals (each from a specific satellite and at a specific frequency) at a rate which is synchronous with the satellite message bit-rate (50 bits per second, or 20 milliseconds per bit). Thus, one complete sequence is completed in a multiple of 20 milliseconds.

- **multiplexing receiver**

A GPS receiver that switches at a very rapid rate between satellites being tracked. Typically, multiplexing receivers require more time for satellite acquisition and are not as accurate as parallel channel receivers. Multiplexing receivers are also more prone to lose a satellite fix in dense woods than parallel channel GPS receivers; GPS receiver that rotates a small number of channels to multiple satellites in order to provide current positioning data. Typically, multiplexing receivers require more time for satellite acquisition and lock on, and are not as accurate as parallel channel receivers. Multiplexing receivers are also more prone to lose satellite fix in dense woods compared to parallel channel GPS receivers.

- **must-carry (retransmission)**

A 1992 Cable Act term requiring a cable system to carry signals of both commercial and noncommercial television broadcast stations that are "local" to the area served by the cable system.

- **nad 27**

Regional horizontal coordinate system used throughout the U.S. until 1986 with the origin at station MEADS RANCH.

- **nad 83**

Horizontal coordinate system for U.S., Canada and Mexico. Originally published in 1986 it is based upon the GRS 80 ellipsoid with its origin at the center of mass defined by BIH at epoch 1984.0. Geodetic surveyors must be particularly cognizant of epoch dates attached to the NAD 83 an acronym. These refer to the mean date of the observations used in the regional adjustment. The California HPGN was originally published at epoch 1991.35.

- **nadir**

  Point on the ground directly in line with the remote sensing system and the center of the earth; A point on the celestial sphere directly below the observer, diametrically opposite the zenith.

- **nagu**

  Notice Advisory to GLONASS Users. A periodic bulletin alerting GLONASS users to changes in system performance.

- **nanometer (nm)**

  One billionth of a meter. Nanometers are used to measure wavelengths in the electromagnetic spectrum.

- **nanu**

  Notice Advisory to NAVSTAR Users. A periodic bulletin alerting GPS users to changes in system performance.

- **nap-of-the-earth flight**

  Flight with a goal to remain close to the earth, usually below the height of surrounding trees and less than 100 ft above the terrain.

- **narrow correlator**

  A correlator in a code-tracking loop in which the spacing between the early and late versions of the receiver-generated reference code is less than one chip. The use of narrow correlators results in pseudorange observations with lower noise.

- **narrow lane**

  The GPS observable obtained by summing the carrier-phase observations simultaneously measured, in cycles, on the L1 and L2 frequencies. The effective wavelength of the narrow-lane observable is 10.7 centimeters. The narrow-lane observable can help resolve carrier-phase ambiguities.

- **narrowcasting**

  Using the electronic media to reach a specific audience

- **nas (network-attached storage)**

  A device used to enlarge the reserve of shared storage space. Also used to accelerate Web page transfer speeds through caching.

- **nasa centers**

  The ten major NASA Centers are:

  1. Ames Research Center (ARC) Located at Moffett Field, California. ARC is active in aeronautical research, life sciences, space science, and technology research. The Center houses the world's largest wind tunnel and the world's most powerful supercomputer system.
  2. The Dryden Flight Research Center, Edwards Air Force Base, California, formerly part of ARC, became a separate entity March 1994. Since the 1940s, this Mojave desert site has been a testing ground for high-performance aircraft

and is one of two prime landing sites for the Space Shuttle.

3. Goddard Space Flight Center (GSFC) Goddard was NASA's first major scientific laboratory devoted entirely to the exploration of space. Located in Greenbelt, Maryland, GSFC's responsibilities include design and construction of new scientific and applications satellites, as well as tracking and communication with existing satellites in orbit. GSFC is the lead center for the Earth Observing System, a key element of Mission to Planet Earth. GSFC also directs operations at the Wallops Flight Facility on Wallops Island, Virginia, which each year launches some 50 scientific missions to sub-orbital altitudes on small sounding rockets.
4. Jet Propulsion Laboratory (JPL) Located in Pasadena, California, JPL is operated under contract to NASA by the California Institute of Technology. Its primary focus is the scientific study of the solar system, including exploration of the planets with automated probes. Most of the lunar and planetary spacecraft of the 1960s and 1970s were developed at JPL. JPL also is the control center for the worldwide Deep Space Network, which tracks all planetary spacecraft.
5. Lyndon B. Johnson Space Center (JSC) Johnson Space Center, located between Houston and Galveston, Texas, is the lead center for NASA's manned space flight program. JSC has been Mission Control for all piloted space flights since 1965, and now manages the Space Shuttle program. JSC's responsibilities include selecting and training astronauts; designing and testing vehicles and other systems for piloted space flight; and planning and executing space flight missions. The center has a major role in developing the Space Station. In addition, JSC directs operations at the White Sands Test Facility in New Mexico, which conducts Shuttle-related tests. The nearby White Sands Missile Range also serves as a backup landing site for the Space Shuttle.
6. Kennedy Space Center (KSC) Located near Cape Canaveral, Florida, KSC is NASA's primary launch site. The Center handles the preparation, integration, checkout, and launch of space

vehicles and their payloads. All piloted space missions since the Mercury program have been launched from here, including Gemini, Apollo, Skylab, and Space Shuttle flights. KSC is the Shuttle's home port, where orbiters are serviced and outfitted between missions, and then assembled into a complete Shuttle "stack" before launch. The Center also manages the testing and launch of unpiloted space vehicles from an array of launch complexes, and conducts research programs in areas of life sciences related to human spaceflight.

7. Langley Research Center (LaRC) Oldest of NASA's field centers, LaRC is located in Hampton, Virginia, and focuses primarily on aeronautical research. Established in 1917 by the National Advisory Committee for Aeronautics, the Center currently devotes two-thirds of its programs to aeronautics, and the rest to space. LaRC researchers use more than 40 wind tunnels to study improved aircraft and spacecraft safety, performance, and efficiency.
8. Lewis Research Center (LeRC) Lewis Research Center, located outside Cleveland, Ohio, conducts a varied program of research in aeronautics and space technology. Aeronautical research includes work on advanced materials and structures for aircraft. Space-related research focuses primarily on power and propulsion. Another significant area of research is in energy and power sources for spacecraft, including the Space Station, for which LeRC is developing the largest space power system ever designed.
9. George C. Marshall Space Flight Center (MSFC). The MSFC, located in Huntsville, Alabama, is responsible for developing spacecraft hardware and systems, and is perhaps best known for its role in building the Saturn rockets that sent astronauts to the Moon during the Apollo program. It is NASA's primary center for space propulsion systems and plays a key role in the development of payloads to be flown on the shuttle (such as Spacelab). MSFC also manages two other NASA sites: the Michoud Assembly Facility in New Orleans where the Shuttle's external tanks are manufactured, and the Slidell Computer Complex in Slidell, Louisiana, which provides computer support to Michoud

and to NASA's John C. Stennis Space Center.

10. John C. Stennis Space Center (SSC). This Center, located on Mississippi's Gulf Coast, is NASA's prime test facility for large liquid propellant rocket engines and propulsion systems. The main mission of the Center is to support testing, on a regular basis, of the Space Shuttle's main propulsion system. SSC is responsible for a variety of research programs in the environmental sciences and the remote-sensing of Earth resources, weather, and oceans, and is the lead NASA Center for the commercialization of space remote sensing.

- **nasa prediction bulletins**

Reports published by NASA's Goddard Space Flight Center providing the latest orbit information on satellites. The report gives information in three parts: 1) two line orbital elements, 2) longitude of the south to north equatorial crossings, and 3) longitude and heights of the satellite crossings for other latitudes.

- **national aerial photography program (napp)**

was a program to photograph the entire United States in infrared and black and white photography. This effort was sponsored by the US Geological Survey.

- **national aeronautics and space administration (nasa)**

U.S. Civilian Space Agency created by Congress. Founded in 1958, NASA belongs to the executive branch of the Federal Government. NASA's mission to plan, direct, and conduct aeronautical and space activities is implemented by NASA Headquarters in Washington, D.C., and by nine major centers spread throughout the United States. Dozens of smaller facilities, from tracking antennas to Space Shuttle landing strips to telescopes are located around the world. The agency administers and maintains these facilities; builds and operates launch pads; trains astronauts; designs aircraft and spacecraft; sends satellites into Earth orbit and beyond; and processes, analyzes, and distributes the resulting data and information. NASA shares responsibility for aviation and space activities with other federal agencies, including the Departments of Commerce, Transportation, and Defense. Much of the work on major projects such as the Space Shuttle and the Space Station is done in the private sector by aerospace companies under government contract.

- **national center for atmospheric research (ncar)**

  Non-profit organization dedicated to furthering understanding of the Earth's atmosphere. Located in Boulder, Co., NCAR is operated by the University Corporation for Atmospheric Research (UCAR) and sponsored by the National Science Foundation (NSF).

- **national differential global positioning system**

  A network of ground- and beacon-based differential stations that increases GPS accuracy.

- **national marine electronics association**

  A US standards committee that defines the data message structure, contents, and protocols that allow GPS receivers to communicate with other electronic equipment.

- **national oceanic and atmospheric administration (noaa)**

  NOAA was established in 1970 within the U.S. Department of Commerce to ensure the safety of the general public from atmospheric phenomena and to provide the public with an understanding of the Earth's environment and resources. NOAA includes: the National Ocean Service which charts the oceans and waters of the U.S. and manages 265,000 acres of estuarine reserves; the National Marine Fisheries Service which maintains the world's largest and most complex marine fisheries management system; the NOAA Corps which operates 18 NOAA research and survey ships and flies 15 NOAA aircraft; and the Office of Oceanic and Atmospheric Research which supports experiments, laboratories, and the National Sea Grant College Program, among other efforts. NOAA has two main components: the National Weather Service (NWS), and the National Environmental Satellite, Data, and Information Service (NESDIS). The National Weather Service provides weather watch and warning services to the public through 57 Weather Service Forecast Offices (WSFO) and over 100 smaller local Weather Service Offices (WSOs) nationwide. Three national forecasting centers provide general and specialized guidance to WSFOs using computer forecast models, satellite data, and conventional surface and upper air observations from around the world. The centers are:

  1. National Meteorological Center, Camp Springs, Maryland;
  2. National Severe Storms Forecast Center, Kansas City, Missouri;
  3. National Hurricane Center, Coral Gables, Florida.

NWS River Forecast Centers (RFCs) provide river stage and flood forecasts. NESDIS provides support to the Weather Service forecast mission by operating a series of environmental satellites and disseminating satellite imagery and derived products to the National Centers and WSFOs. NESDIS operates three national data and information centers: the National Geophysical Data Center, the National Climatic Data Center (NCDC), and the National Oceanographic Data Center (NODC). NOAA organizations perform numerous services in addition to monitoring weather conditions. They assess crop growth and other agricultural conditions, sense shifting ocean currents, and measure surface temperatures of oceans and land. They relay data from surface instruments that sense tide conditions, Earth tremors, river levels, and precipitation.

- **national space science data center (nssdc)**

The NSSDC provides on-line and off-line access to a wide variety of astrophysics, space plasma and solar physics, lunar and planetary, and Earth science data from NASA space flight missions, in addition to selected other data, models, and software. Located at Goddard Space Flight Center (GSFC) in Greenbelt, Maryland, the NSSDC is sponsored by the Information Systems Office of NASA's Office of Space Sciences. NSSDC on-line data and services are currently free of charge, off-line support (e.g., replications and mailing of magnetic tapes) are available for the cost of fulfilling the request. The NSSDC Master Catalog (NMC) provides an on-line listing of available data sets and the forms that the data are available in (such as CD-ROM), and provides information about the spacecraft and experiments (including past, present, and future NASA and non NASA) from which these data were obtained. The on-line NASA Master Directory (NMD) identifies and briefly describes data of potential interest to the NASA research community, and where possible, provides electronic links to publicly accessible data at sites world-wide. On-line information services are made available through the menu-based NSSDC Online Data Information Service (NODIS).

- **national transportation communication for intelligent transportation systems protocol (ntcip)**

Provides a communication standard that ensures the

interchangeability of traffic control and ITS devices.

- **nautical mile**
A unit of distance (U.S.) equal to exactly 1.852 kilometers or about 6076.1 feet. A nautical mile is approximately equal to 1/60 of a degree or 1 minute of arc of a great circle of the Earth (i.e., 1 minute of arc of latitude or of longitude at the equator); A distance of 1851 9999 meters which is one minute of arc of a great circle of the Earth (1.15 times longer than a statute mile, or 6076.1 feet).

- **navd 88**
Vertical (elevation) reference system for U.S., Canada and Mexico. Published in 1991 the orthometric heights are derived from an adjustment of leveling data constrained at a single point on the St. Lawrence Seaway.

- **navdata**
A 1500-bit navigation message broadcast at 50 bits per second on the L1 and L2 channels. The message contains time, clock correction, ephemeris, satellite health and ionospheric delay model parameters.

- **navigation**
The process of directing a vehicle so as to reach the intended destination; A system, usually software, in which the primary purpose is to generate position relative to a coordinate frame, usually fixed earth frame, such as latitude and longitude or UTM;

- **navigation coordinate**
Any one of a set of quantities; the set serving to define a position.

- **navigation message**
The message transmitted by each GPS satellite containing system time, clock correction parameters, ionospheric delay model parameters, and the satellite's ephemeris data and health. The information is used to process GPS signals to give the user time, position, and velocity. Also known as the data message; A 37500-bit data message included in the GPS signal. The message, sent at a rate of 50 bits per second, includes the satellite ephemeris, clock data, almanac, and other information about the satellites and their signals; Navigation data is transmitted at 50 bits per second and contain ephemeris and clock data for that particular satellite, almanac data for all GPS satellites, and other information. The data is transmitted in 1500 bit frames, thus taking 30 seconds to transmit. A complete set of navigation data requires 12-1/2 minutes to transmit. Similar messages are sent by GLONASS satellites.

- **navigation parameter**
A measurable characteristic of motion or position used in the process of navigation.

- **navigation quantity**

A measured value of a navigation parameter.

- **navigation reference point (nrp)**

A point, usually fixed in earth coordinates but possibly moving; Also, a basic guidance mode, providing lateral guidance to an NRP, either by course or by direct (operator selectable).

- **navigation satellite for timing and ranging (navstar)**

NAVSTAR is the original designation given to the Global Positioning System (GPS) by the U.S. Department of Defense. NAVSTAR is occasionally used today, either by itself or in conjunction with GPS, as in NAVSTAR GPS.

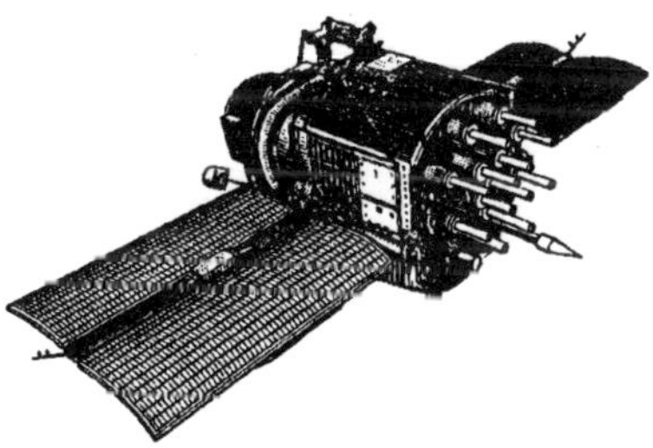

- **navigation solution**

Complex of all the navigation coordinates needed to define a position. See: final navigation solution.

- **navigational aid**

An instrument, system, device, chart, or method intended to assist in the navigation of a vehicle.

- **navtech cds**

NAVTECH makes several mapping discs, each containing detailed maps of different regions of the U.S. Each disc contain all interstate highways, federal/state highways, and county roads within a region, as well as street information for cities and larger towns.

- **ndgps**

Nationwide Differential GPS. The NDGPS plan calls for the conversion of a number of U.S. Air Force Ground Wave Emergency Network (GWEN) sites to broadcast DGPS signal. The range is 250 miles.

- **near infrared (nir)**

The shorter wavelength range of the infrared region of the EM spectrum, from 0.7 to 2.5 m. It is often divided into very-near infrared (VNIR) covering the range accessible to photographic emulsions (0.7 to 1.0m), and the short-wavelength infrared (SWIR) covering the remainder of the NOR atmospheric window from 1.0 to 2.5m.

- **near range**

Refers to the portion of a radar image closest to the aircraft or satellite flight path.

- **neatline**
  Line separating the body of a map from the map margin. On a standard quadrangle map, the neatlines are the meridians and parallels delimiting the quadrangle.

- **negative photograph**
  Photograph on film or paper in which the relationship between bright and dark tones is the reverse of that of the features on the terrain.

- **nema**
  National Electrical Manufacturers Association.

- **nema 0183**
  A communication protocol used by GPS units and other types of navigation and marine electronics.

- **nephanalysis**
  A type of analysis using satellite cloud pictures to study the relationship between cloud forms and storm systems. In classical mythology, Nephele was a woman Zeus formed from a cloud.

- **nepheloccygia**
  Clouds that resemble recognizable shapes.

- **network**
  Any connection of two or more computers that enables them to communicate. Networks may include transmission devices, servers, cables, routers and satellites. The phone network is the total infrastructure for transmitting phone messages.

- **network adjustment**
  A form of least squares solution in which the observed baseline vectors are treated as "observations" in a secondary adjustment. It may be a minimally constrained network adjustment with only one station coordinate held fixed, or it may be constrained by more than one fixed (known) coordinates. The latter is typical of a GPS survey carried out to densify or connect some newly coordinated points to a previously established control or geodetic framework.

- **newton's law of universal gravitation**
  All bodies attract each other with what is called gravitational attraction. This applies to the largest stars as well as the smallest particles of matter. The force of attraction between two small bodies (or between two spherical bodies of any size) is proportional to the product of their masses and inversely proportional to the square of the distance between their centers. In other words, the closer two bodies are to each other, the greater their mutual attraction. As a result, to stay in orbit, a satellite needs more speed in a low than a high orbit.

- **newton's laws of motion**

  Newton's three laws of motion are:

  1. Every body continues in a state of uniform motion in a straight line unless acted upon by some external force.
  2. The time rate of change of momentum (mass x velocity) is proportional to the impressed force. In the usual case where the mass does not change, this law can be expressed in the familiar form: force = mass x acceleration or F = ma.
  3. To every force or action, there is always an equal and opposite reaction.

- **ngdc**

  National Geophysical Data Center, located in Boulder, Colorado. See National Oceanic and Atmospheric Administration.

- **ngs-national geodetic survey**

  An agency of the United States NOAA that defines and manages a national coordinate system. This network, the National Spatial Reference System (NSRS), provides the foundation for transportation and communication; mapping and charting; and a multitude of scientific and engineering applications.

- **ngvd 29**

  Mean sea level reference system for U.S. prior to NAVD 88. Constrained to 26 tide stations in U.S. and Canada, the datum is distorted to fit sea surface topography.

- **nibble**

  Four bits of data.

- **nima**

  National Imaging and Mapping Agency.

- **nimbus satellite program**

  A NASA program to develop observation systems meeting the research and development requirements of atmospheric and Earth scientists. The Nimbus satellites, first launched in 1964, carried a number of instruments: microwave radiometers, atmospheric sounders, ozone mappers, the Coastal Zone Color Scanner (CZCS), infrared radiometers, etc. Nimbus-7, the last in the series, provided significant global data on sea-ice coverage, atmospheric temperature, atmospheric chemistry (i.e. ozone distribution), the Earth's radiation budget, and sea-surface temperature.

- **nmea**

  National Marine Electronics Association, a U.S. standards body that defines message structure, content and protocols to allow electronic equipment installed within ships and boats to communicate with each other. GPS receivers can be configured to output various types of messages in the "NMEA format"; A U.S. standards committee that defines data message structure, contents, and protocols to allow the GPS receiver to communicate with other pieces of electronic equipment aboard ships.

- **nmea 0183**
  The committee of the National Marine Electronics Association that developed a standard for interfacing marine electronic devices. The standard is widely used for interfacing GPS receivers. nmea national marine electronics association. A professional organization that defines and maintains the standard serial format used by marine electronic navigation equipment and computer interfaces.

- **nmea standard**
  A NMEA standard defines an electrical interface and data protocol for communications between marine instrumentation.

- **nodal period**
  The amount of time difference between two successive ascending nodes of a satellite orbit.

- **noise**
  Part of received data that is undesired, consisting of random sinusoidal terms added to a signal; Random or repetitive events that obscure or interfere with the desired information; An interfering signal that tends to mask the desired signal at the receiver output and which can be caused by space and atmospheric phenomena, can be human made, or can be caused by receiver circuitry.

- **noise figure (nf)**
  A term which is a figure of merit of a device, such as an LNA or receiver, expressed in dB, which compares the device with a perfect device.

- **nondirectional filter**
  Mathematical filter that treats all orientations of linear features equally.

- **non-selective scattering**
  The scattering of EM energy by particles in the atmosphere which are much larger than the wavelengths of the energy, and which causes all wavelengths to be scattered equally.

- **non-spectral hue**
  A hue which is not present in the spectrum of colours produced by the analysis of white light by a prism of diffraction grating. Examples are brown, magenta, and pastel shades.

- **nonsystematic distortion**
  Geometric irregularities on images that are not constant and cannot be predicted from the characteristics of the imaging system.

- **non-volatile memory**
A type of memory device that retains data in the absence of a power supply.

- **normal color film**
Film in which the colors are essentially true representations of the colors of the terrain.

- **normalizer**
Function to restrict input to a specific range, such as restricting an angle alpha in radians so that -pi le alpha le +pi; angles usually require normalizing following any computation; normalized variables often present problems for filters and other functions at their discontinuities.

- **north up orientation**
Fixes the GPS receiver's map display so north is always fixed at the top of the screen.

- **northing**
A term used to denote positions north of the origin in the UTM coordinate grid system.

- **notice advisory to glonass users (nagu)**
A periodic bulletin alerting GLONASS users to changes in system performance.

- **notice advisory to navstar users (nanu)**
A periodic bulletin alerting GPS users to changes in system performance.

- **nsdi-national spatial data infrastructure**
The set of base geographic data necessary for effective operation of the United States federal government and its suppliers, made accessible as a distributed database.

- **nsrs-national spatial reference system**
A system administered by the NGS that provides the foundation for transportation and communication; mapping and charting; and a multitude of scientific and engineering applications.

- **ntsc - national television standards committee**
A video standard established by the United States (RCA/NBC} and adopted by numerous other countries. This is a 525-line video with 3.58-MHz chroma subcarrier and 60 cycles per second; The organization (RCA/NBC) which established the standards for the television broadcast system in the United States. A close approximation of this system was also adopted in Japan. The US and Japanese video systems are now referred to as 'NTSC' systems. This video standard provides for a screen density of 525 scanlines per picture and operates at 60 cycles per second. This system was developed so that existing black and white (B/W) mono audio

television receivers would be compatible with color stereo audio broadcasts.

- **null field**
By NMEA standard, indicates that data is not available for the field. Indicated by two ASCII commas, i.e., ",," (HEX 2C2C), or, for the last data field in a sentence, one comma followed by either the checksum delimiter "*" (HEX 2A) or the sentence delimiters <CR><LF> (HEX 0D0A).

- **number portability**
A term used to describe the capability of individuals, businesses and organizations to retain their existing telephone number(s) -- and the same quality of service -- when switching to a new local service provider.

- **oasis**
A spot in a desert made fertile by water, which normally originates as groundwater.

- **oblique photograph**
Photograph acquired with the camera intentionally directed at some angle between horizontal and vertical orientations.

- **obscuration**
Term used to describe periods of time when a GPS receiver's line-of-sight to GPS satellites is blocked by natural or man-made objects.

- **observation**
Any measurement. The two observations used in NovAtel's RTK algorithms are the pseudorange measurement and the carrier phase measurement.

- **observation set**
A set of GPSCard measurements taken at a given time which includes one time for all measurements, and the following for each satellite tracked: PRN number, pseudorange or carrier phase or both, lock time count, signal strength, and tracking status. Either L1 only or L1 and L2 measurements are included in the set. The observation set is assumed to contain information indicating how many satellites it contains and which ones have L1-only and which ones have L1/L2 pairs.

- **observing session**
The period of time over which GPS data are collected simultaneously by two or more receivers.

- **obstacle avoidance (oa)**
Flight cues designed to avoid obstacles, such as terrain, buildings, and power lines; Compare: terrain avoidance, threat avoidance.

- **occluded front (occlusion)**
A composite of two fronts formed as a cold front overtakes a warm front. A cold occlusion results when the coldest air is behind the cold front. The cold front undercuts the warm front and, at

the Earth's surface, coldest air replaces less-cold air. A warm occlusion occurs when the coldest air lies ahead of the warm front. Because the cold front can not lift the colder air mass, it rides piggyback up on the warm front over the coldest air.

- **ocean**
  The salt water surrounding the great land masses. The land masses divide the ocean into several distinct portions, each of which also is called an ocean. The oceans include the Pacific Ocean, the Atlantic Ocean, the Indian Ocean, and the Arctic Ocean.

- **oceanic survey**
  Survey or examination of condition in the ocean or any part of it, with reference to animal or plant life, chemical elements present, temperature gradients, etc.

- **oc-n (optical carrier level-n)**
  The optical equivalent of (STS-N) Typical OC Levels are:
  OC-1: 51.84 Mb/s.
  OC-3: 155.52 Mb/s.
  OC-12: 622.08 Mb/s.
  OC-24: 1.244 Gb/s.
  OC-48: 2.488 Gb/s.
  OC-96: 4.976 Gb/s.

- **oem**
  Original Equipment Manufacturer. Typically GPS receiver "boardsets" or "engines" that a product developer can embed within some application or hardware package.

- **off**
  Device is powered off (power switch is off; no response to communications) - no data and function is available.

- **offset**
  Part of received data that is undesired, consisting of a random, time-invariant term added to a signal; Synonyms: bias.

- **omega**
  An OMEGA receiver provides position information that can be used to update the aircraft navigation position. OMEGA is the least accurate method of obtaining position information.

- **omnistar**
  A wide-area GPS correction service, using L-band satellite broadcast frequencies (1525 - 1560 MHz). Data from many widely-spaced Reference Stations is used in a proprietary multi-site solution. OmniSTAR Virtual Base Station (VBS) types acheive sub-meter positioning over most land areas worldwide while OmniSTAR High Performance (HP) types achieve 10 cm accuracy. Use of OmniSTAR service requires a subscription.

- **on the fly (otf)**
  OTF is a common term used to describe the technique of resolving differential carrier-

phase integer ambiguities without requiring a global navigation satellite receiver to remain stationary; This is a form of Ambiguity Resolution (AR) which does not require that the receivers remain stationary for any length of time. Hence this AR technique is suitable for initialising carrier phase-based Kinematic Positioning. For many applications this introduces considerable flexibility. For example, aircraft do not have to be parked on the ground in order to resolve the carrier cycle ambiguities, and then require that signal lock-on be maintained throughout the kinematic survey. However, dual-frequency instrumentation capable of making both carrier phase and precise (P-Code level) pseudo-range measurements is required.

- **on-line**

Electronic availability on demand from a computer-based system without mounting removable media such as magnetic tape or disks.

- **open video systems**

An alternative method to provide cable-like video service to subscribers.

- **operations**

How well is equipment operating; Values: operational (all function and data is available), degraded (equipment has partially failed with some function or data unavailable and some available), failed (equipment has failed with no function or data available);

- **operator service provider (osp)**

A common carrier that provides services from public phones, including payphones and those in hotels/motels.

- **orbit**

The path described by a heavenly body in its periodic revolution. Earth satellite orbits with inclinations near 0 degree are called equatorial orbits because the satellite stays nearly over the equator. Orbits with inclinations near 90 degrees are called polar orbits because the satellite crosses over (or nearly over) the north and south poles. See orbital inclination.

- **orbital period**

The time that it takes a satellite to complete one circumnavigation of its orbit.

- **orbital plane**

An imaginary plane, extending throughout space, that contains the satellite orbit

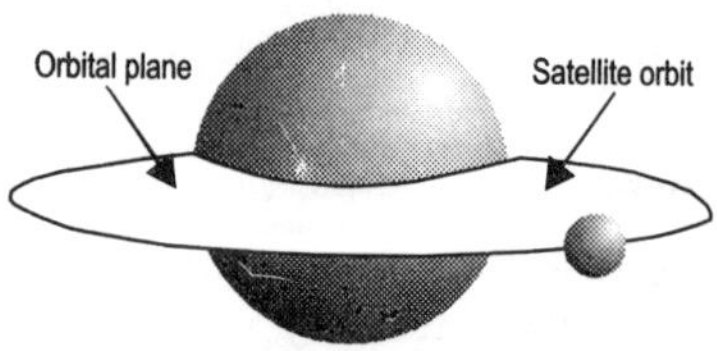

An imaginary gigantic flat plate containing an Earth satellite's orbit.

The orbital plane passes through the center of the Earth.

- **order of survey**

In an analogous manner to "Class of Survey", Order of Survey is a means of categorising the quality, or precision, of a static survey. However, it relates to the external quality, and is influenced by the quality of the "external" network information. The number of categories, the notation applied, and the accuracy tolerances defining the transition from one order to another are defined by individual nations. Typically they mirror the categories of Class of Survey, hence an A Class survey may correspond to a 1st Order survey. The labeling of a particular Order (e.g. 1st, 2nd, etc.) to a survey points within a "network" (whether it is carried out using GPS or any other technique) is performed as part of the process of Network Adjustment in which the relative error ellipses (in the horizontal case) between coordinated stations are computed and compared with the accuracy standards that must be met for various categories of Order. However, unlike the Minimally Constrained Network Adjustment that is a prerequisite to establishing the Class of Survey, the Network Adjustment must be constrained to the surrounding geodetic control. Hence a very high quality GPS network (therefore a high Class survey) may be distorted to "fit" the existing control which may have been determined using a lower Class survey. The resulting Order of the Survey would have to match the lower of either the Class of the GPS survey or the Class of the existing geodetic control. If the existing geodetic control is of a lower quality to what can be achieved using modern GPS Surveying techniques, then the geodetic control network must be upgraded or "renovated" using more precise GPS Geodesy techniques.

- **oregon's gps technology**

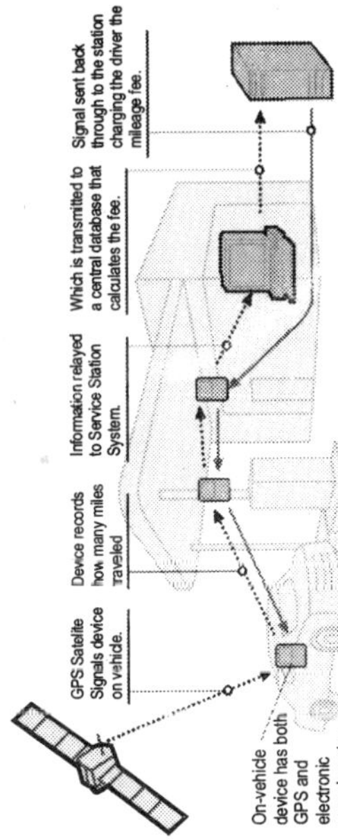

Oregon is experimenting with technology that would allow it to measure how many road miles a vehicle travels. A global positioning system, combined with an electronic speedometer would work in concert to measure only miles traveled in Oregon. The driver would then be taxed based on miles traveled, rather than by gallons of gasoline.

- **origin waypoint**
  The starting point of the present navigation leg, expressed in latitude and longitude.

- **original equipment manufacturer (oem)**
  They are manufacturers of equipment designed for use by end consumers. OEM GPS modules are named as such because they require electronics knowledge to interface and therefore are used most commonly by OEMs.

- **orthographic projection**
  The two-dimensional graphic representation of an object formed by the perpendicular intersections of lines drawn from points on the object to a plane of projection.

- **orthometric height elevation**
  The distance from the geoid to a point, measured along a line normal to the geoid.

- **orthophotograph**
  A vertical aerial photograph from which the distortions due to varying elevation, tilt, and surface topography have been removed, so that it represents every object as if viewed directly from above; Photograph having the properties of an orthographic projection. It is derived from a conventional perspective photograph by simple or differential rectification so that image displacements caused by camera tilt and terrain relief are removed.

- **orthophotomap**
  Orthophotographic map with contours and cartographic features related to standard reference systems.

- **orthophotoquad**
  Monocolor orthophotgraphic map presented in a standard quadrangle format and related to standard reference systems. It has no contours and little or cartographic treatment.

- **orthophotoscope**
  An optical-electronic device which converts a normal vertical aerial photograph to an orthophotograph; Photomechanical device used in conjunction with a double-projection stereoplotter for producing an orthophotograph.

- **osa (open systems architecture)**
  Permits transparent communication between cooperating machines. OSA is embodied in international standards referred to as the Open Systems Interconnection Reference Model (OSI).

- **oscillator**

A device that generates a signal of a given frequency.

- **osgb**

A coordinate system describing only Great Britain, similar to UTMs. Used with GRB36 datum.

- **osgb-ordnance survey of great britain**

A system similar to the U.S. UTM for Great Britain.

- **osi (open systems interconnection reference model)**

The reference model, also referred to as OSIRM, provides a framework for communication that relies on standard non-proprietary interconnection protocols (protocol stack). It consists of seven layers that serve to decompose the complexity of information flow between cooperating machines into consecutive steps that are substantially independent of each other. The seven OSI model layers are:

Layer #7 Application (User): Handles issues like network transparency, resource allocation and problem partitioning. The application layer is concerned with the user's view of the network (e.g. formatting electronic mail messages). The presentation layer provides the application layer with a familiar local representation of data independent of the format used on the network.

Layer #6 Presentation: Performs functions such as text compression, code or format conversion to try to smooth out differences between hosts. Allows incompatible processes in the application layer to communicate via the session layer.

Layer #5 Session: Uses the transport layer to establish a connection between processes on different hosts. It handles security and creation of the session. It is used by the presentation layer.

Layer #4 Transport: Determines how to use the network layer to provide a virtual error-free, point to point connection so that host A can send messages to host B and they will arrive un-corrupted and in the correct order. It establishes and dissolves connections between hosts. It is used by the session layer.

Layer #3 Network: Determines routing of packets or data from sender to receiver via the data link layer and is used by the transport layer. The most common network layer protocol is IP.

Layer #2 Data Link: Splits data into frames for sending on the physical layer and receives acknowledgment frames. It performs error checking and re-transmits frames not received correctly. It provides an error-free virtual channel to the network layer.

Layer #1 Physical: Concerning electrical and mechanical connections to the network. Examples of physical layer protocols are CSMA/CD, token ring and bus. Note: Compare with the layers of the TCP/IP protocol suite.

- **osm**

  Ordnance Survey Maps. Britain's National Mapping Agency

- **outage**

  Defined as a loss of Availability, due to either there not being enough satellites visible to calculate a position, or the value of the DOP indicator is greater than some specified value (implying that the accuracy of the position is unreliable).

- **outer marker**

  Marker beacon located 5-7mi from the end of the runway.

- **output power**

  The amount of electrical energy applied to the transducer by the sonar transmitter. Expressed in watts, typically the higher the output power the sonar unit has, the deeper it can penetrate, and the finer the detail that can be shown.

- **overedge**

  Any portion of a map lying outside the nominal map border (neatline).

- **overlap**

  Extent to which adjacent images or photographs cover the same terrain, expressed as a percentage.

- **override**

  To alter selection made automatically by software

- **oziexplorer**

  Scanned map software for use with GPS.

- **ozone**

  An almost colorless, gaseous form of oxygen with an odor similar to weak chlorine.

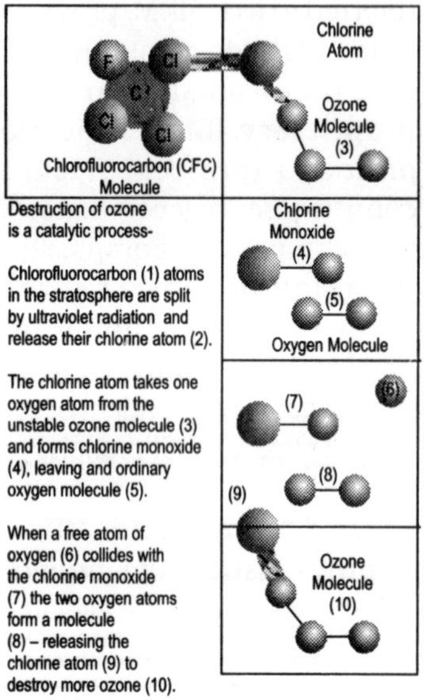

A relatively unstable compound of three atoms of oxygen, ozone constitutes--on the average--less than one part per million (ppm) of the gases in the atmosphere (peak ozone concentration in the stratosphere can get as high as 10 ppm). Yet ozone in the stratosphere absorbs nearly all of the biologically damaging solar ultraviolet radiation before it reaches the Earth's surface where it can cause skin cancer, cataracts, and immune deficiencies, and can harm crops and aquatic ecosystems. Ozone is produced naturally in the middle and upper stratosphere through dissociation of molecular oxygen by sunlight. In the absence of chemical species produced by human activity, a number of competing chemical reactions among naturally occurring species--primarily atomic oxygen, molecular oxygen, and oxides of hydrogen and nitrogen--maintains the proper ozone balance. In the present-day stratosphere, this natural balance has been altered, particularly by the introduction of man-made chlorofluorocarbons. If the ozone decreases, the ultraviolet radiation at the Earth's surface will increase. Tropospheric ozone is a by-product of the photochemical (light-induced) processes associated with air pollution. See photochemical smog. Ozone in the troposphere can damage plants and humans.

- **ozone hole**

A large area of intense stratospheric ozone depletion. over the Antarctic continent that typically occurs annually between late August and early October, and generally ends in mid-November. This severe ozone thinning has increased conspicuously since the late seventies and early eighties. This phenomenon is the result of chemical mechanisms initiated by man-made chlorofluorocarbons. Continued buildup of CFCs is expected to lead to additional ozone loss worldwide. The thinning is focused in the Antarctic because of particular meteorological conditions there.

- **ozone layer**

The layer of ozone that begins approximately 15 km above Earth and thins to an almost negligible amount at about 50 km, shields the Earth from harmful ultraviolet radiation from the sun.

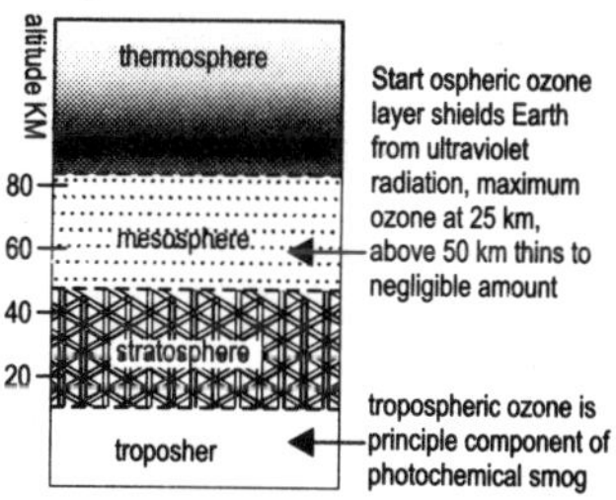

The highest natural concentration of ozone (approximately 10 parts per million by volume) occurs in the stratosphere at approximately 25 km above Earth. The stratospheric ozone concentration changes throughout the year as stratospheric circulation changes with the seasons. Natural events such as volcanoes and solar flares can produce changes in ozone concentration, but man-made changes are of the greatest concern.

- **ozone mini-hole(s)**

Rapid, transient, polar-ozone depletion. These depletions, which take place over a 50-kilometer squared area, are caused by weather patterns in the upper troposphere. The decrease in ozone during a mini-hole event is caused by transport, with no chemical depletion of ozone. However, the cold stratospheric temperatures associated with weather systems can cause clouds to form that can lead to the conversion of chlorine compound from inert to reactive forms. These chlorine compounds can then produce longer-term ozone reductions after the mini-hole has passed.

- **ozone-measuring satellite instruments**

Satellite-based ozone-measuring instruments can measure ozone by looking at the amount of ultraviolet absorption reflected from the Earth's surface and clouds. Some instruments provide data within the different levels of the atmosphere. The Total Ozone Mapping Spectrometer (TOMS) maps the total amount of ozone between ground and the top of the atmosphere. The amount and distribution of ozone molecules in the stratosphere varies greatly over the globe, changing in response to natural cycles such as seasons, sun cycles, and winds. Utilizing satellites has enabled scientists to assess ozone levels simultaneously over the entire Earth, and has led them to conclude that global ozone levels are being depleted.

- **p code**

The Precise or Protected code. A very long sequence of pseudo-random binary biphase modulations on the GPS carrier at a chip rate of 10.23 MHz, which repeats about every 267 days. Each 1-week segment of this code is unique to one GPS satellite and is reset each week

- **packet**

A single group of digital signals sent from an originator to an addressee; A unit of data that is routed between an origin and a destination on the Internet or any other packet-switched network; A sequence of bits that is divided into two parts -- one contains the user's information, and the other contains control information; In data communication, a grouping of a sequence of binary digits, including data and control signals, that is transmitted and switched as a composite whole. The data, control signals, and possibly error control information, are arranged in a specific format. The packet can be of either fixed or variable length.

- **packet loss**
  Packet Loss and Packet Jitter are types of degradation in the delivery stream of digital information which result from unfavorable network conditions such as extremely heavy or bursty traffic. Individual packets comprising the data stream can arrive early, late or out of sequence causing unstable video reception, pauses, the appearance of artifacts or the loss of the data altogether.

- **packet messaging**
  Radio transmitted message system.

- **packet mode**
  A packet switched operational mode for transferring (transporting and switching) user information through a network without establishing a connection. The packets do not necessarily arrive at their destination in the order they were sent, unlike the circuit mode of transmission. See packet switching.

- **packet radio**
  The radio transmission of data in packets between stations.

- **packet switching**
  Data transmission method that divides messages into standard-sized packets for greater efficiency of routing and transport through a network.

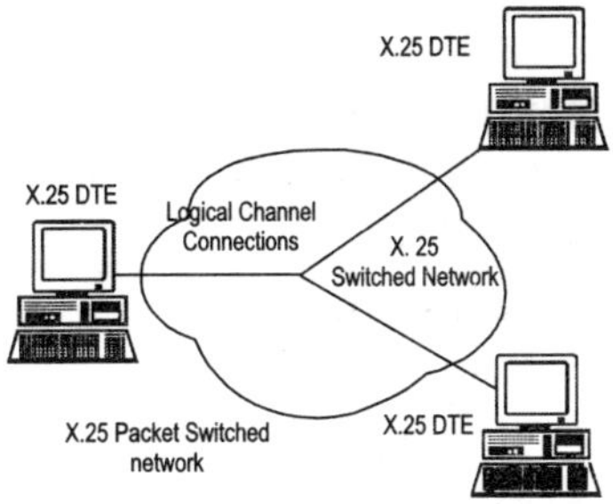

- **paging system**
  A one-way mobile radio service where a user carries a small, lightweight miniature radio receiver capable of responding to coded signals. These devices, called "pagers," emit an audible signal, vibrate or do both when activated by an incoming message.

- **pal - phase alternation system**
  The German developed TV standard based upon 50 cycles per second and 625 lines; A German developed color television standard which provides for a screen density of 625 scanlines per picture and which operates at 50 cycles per second. This system is common in the UK, much of Europe, Africa, Asia and Oceania and is incompatible with the North American and Japanese NTSC

standards. The IntelSat satellite system often employs the PAL standard.

- **paleoclimate**
  Climate as it existed in the distant past, particularly before historical records.

- **paleogeography**
  The study of ancient or prehistoric geography.

- **panchromatic**
  Sensitive to all or most of the visible spectrum.

- **panchromatic film**
  Black and white film that is sensitive to all visible wavelengths.

- **pantrack™**
  A Garmin-proprietary feature that allows the user to move the pointer and pan a track in either direction, then select a location along the track to start a TracBack® or GoTo, or to mark a waypoint.

- **parabolic antenna**
  The most frequently found satellite TV antenna, it takes its name from the shape of the dish described mathematically as a parabola. The function of the parabolic shape is to focus the weak microwave signal hitting the surface of the dish into a single focal point in front of the dish. It is at this point that the feed horn is usually located.

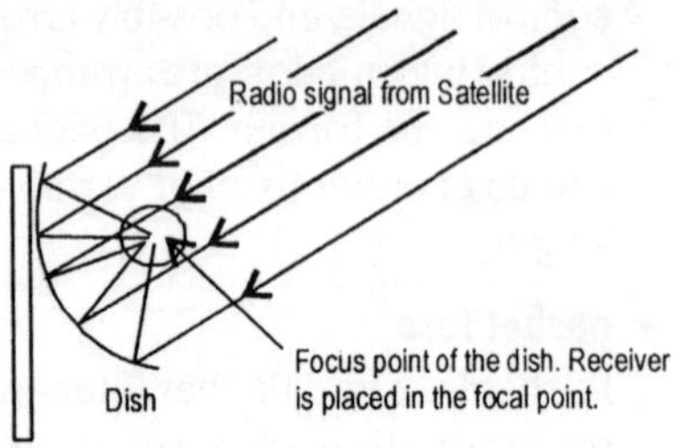

- **parallax**
  Displacement of the position of a target in an image caused by a shift in the observation system.

- **parallax difference**
  The difference in the distance on overlapping vertical photographs between two points, which represent two locations on the ground with different elevations.

- **parallel**
  A line of latitude.

- **parallel channel receiver**
  A continuous tracking receiver using multiple receiver circuits to track more than one satellite simultaneously. GPS receiver that simultaneously tracks multiple satellites to provide the fastest, most reliable and accurate navigational data...under the severest of environmental conditions.

- **parallel of latitude**
  An imaginary line circling the surface of the earth that is parallel to the equator; a line of latitude.

- **parallel receiver**
A receiver that monitors four or more satellites simultaneously with independent channels.

- **parallel-polarized**
Describes a radar pulse in which the polarization of the return is the same as that of the transmission. Parallel-polarized images may be HH (horizontal transmit, horizontal return) or VV (vertical transmit, vertical return).

- **paramo**
A treeless alpine plateau of the Andes and tropical South America.

- **parity**
The even or odd quality of the number of ones or zeroes in a binary code. Parity is often used to determine the integrity of data especially after transmission.

- **parity error**
A digital message is composed of 1s and 0s. Parity can be defined as the sum of these bits within a word unit. A parity error results when one of the bits is changed so that the parity calculated at reception is not the same as it was at transmission of the message.

- **pascal (pa)**
Unit of atmospheric pressure named in honor of Blaise Pascal (1632-1662), whose experiments greatly increased knowledge of the atmosphere. A Pascal is the force of one Newton acting on a surface area of one square meter. It is the unit of pressure designated by the International System.

- **pass**
In digital filters, refers to the spatial frequency of data transmitted by the filter. High-pass filters transmit high-frequency data; low-pass filters transmit low-frequency data.

- **passband**
The range of frequencies handled by a satellite translator or transponder.

- **passive microwaves**
Radiation in the 1 mm to 1 m range emitted naturally by all materials above absolute zero.

- **passive remote sensing**
Remote sensing of energy naturally reflected or radiated from the terrain.

- **passive system**
A system sensing only radiation emitted by the object being viewed or reflected by the object from a source other than the system. See active system.

- **pass-to-pass accuracy**
Also called repeatability. In the context of parallel tracking, this is how far off you are from one path to the next, regardless on which pass you are on. Mostly the pass-to-pass accuracy is aproximately 2 to 3 times more precise than the statistic accuracy.

- **past status words**

  Status words that are logically combined over time (such as "and" ing or "or" ing) to provide history of what has been set in the past.

- **patch antenna**

  An antenna based on a plated section of metal on a circuit board and its associated electronics used for receiving GPS signals.

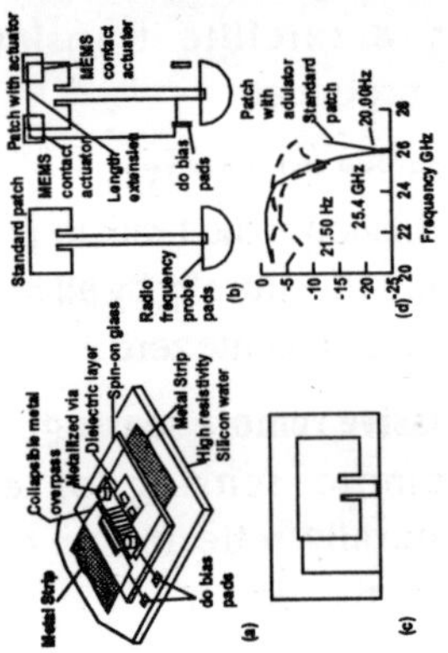

- **path**

  A line connecting a series of points in space and constituting a proposed or traveled route. See: course line.

- **path integral gain (kint)**

  A guidance control law parameter, generated by the lateral guidance modes;

- **path integral value (intval)**

  A guidance control law parameter, generated by the lateral guidance modes; Typical Units: ft; Dimensions: Length

- **path-and-row index**

  System for locating Landsat MSS and TM images.

- **pattern**

  Regular repetition of tonal variations on an image or photograph.

- **payload**

  The instruments that are accommodated on a spacecraft.

- **pcmcia-personal computer memory card international association**

  A standard for a credit card-size memory or input/output device that fits into a notebook, laptop or personal computer.

- **p-code**

  Precise code or protected code. A pseudorandom string of bits that is used by GPS receivers to determine the range to the transmitting GPS satellite. P-code is replaced by an encrypted Y-code when Anti-Spoofing is active. Y-code is intended to be available only to authorized (primarily military) users.

- **pda (personal digital assistant)**

  Small, hand-held devices offering functions similar to a personal computer (though limited); also now offering access to some ITS services.

- **pdop**

  Position Dilution of Precision, a unitless figure of merit expressing the relationship between the error in user position and the error in satellite position. Geometrically, PDOP is proportional to 1 divided by the volume of the pyramid formed by lines running from the receiver to four satellites observed. Values considered "good" for positioning are small, say 3. Values greater than 7 are considered poor. Thus small PDOP is associated with widely separated satellites. PDOP is related to horizontal and vertical DOP by PDOP2 = HDOP2 + VDOP2. Small PDOP is important in positioning, but much less so in surveying.

- **peak-to-peak**
  Normally used to describe the output power of a sonar transmitter. This value is a measurement of the total swing of an AC voltage from its peak negative value to its peak positive value.

- **pedal**
  A flight control operated by pushing with feet, primarily to control yaw via the rudder in fixed-wing aircraft or thrust to tail rotor in rotary-wing aircraft; pedals are automatically controlled in modern aircraft;

- **perigee**
  The point in a body's orbit at which it is nearest the earth.

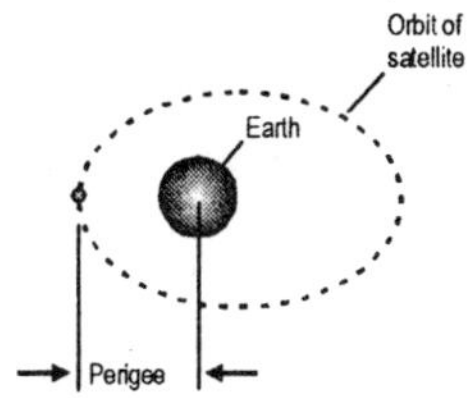

- **perihelion**
  The point in the orbit of a planet or comet which is nearest the Sun (as opposed to the aphelion, which is the point in the orbit farthest from the Sun).

- **period**
  Time of a periodic process; 1/f where f is the sampling frequency; Symbols: T; Typical Units: s; Dimensions: Time; The time required for a satellite to make one complete revolution about the Earth (see Anomalistic Period and Nodal Period).

- **period decay (aka decay)**
  The tendency of a satellite to lose orbital velocity due to the influence of atmospheric drag and gravitational forces. A decaying object eventually impacts the surface of the Earth or burns up in the atmosphere. This parameter directly affects the satellite's mean motion.

- **periodic built-in-test (pbit)**
  Self tests running internal to a device as part of normal operation.

- **periodic line dropout**
  Defect on Landsat MSS or TM images in which no data are

recorded for every sixth or sixteenth scan line, causing a black line on the image.

- **periodic line striping**

Defect on Landsat MSS or TM images in which every sixth or sixteenth scan line is brighter or darker than the others. Caused by the sensitivity of one detector being higher or lower than the others.

- **personal communications service (pcs)**

Any of several types of wireless, voice and/or data communications systems, typically incorporating digital technology. PCS licenses are most often used to provide services similar to advanced cellular mobile or paging services. However, PCS can also be used to provide other wireless communications services, including services that allow people to place and receive communications while away from their home or office, as well as wireless communications to homes, office buildings and other fixed locations.

- **personnel locating system (pls)**

A basic guidance mode, providing lateral guidance to a PLS transmitter from range and bearing inputs. Equipment that determines range and bearing to a personnel with a PLS transmitter; Provides range and bearing to locate ground personnel. Coded continuous or periodic interrogations of the portable ground radios are used to provide the information.

- **perturbations**

Minor corrections to the Keplerian model of a satellite orbit as an ellipse of constant shape and orientation. Since satellite orbits are affected by Earth's gravity and drag caused by the Earth's atmosphere (causing satellites to spiral downward), minor adjustments must be made to the orbit.

- **pH**

A symbol for the degree of acidity or alkalinity of a solution. Expressed as a negative logarithm of the hydrogen ion concentration in a solution, pH = -log10[H+]. If the hydrogen ion concentration of a solution increases, the pH will decrease, and vice versa. The value for pure distilled water is regarded as neutral, pH values from 0 to 7 indicate acidity, and from 7 to 14 indicate alkalinity.

- **phase angle**

The time difference between the same point on two different waves, usually measured in fractions of a cycle (radians or degrees)

- **phase center**

The apparent center of signal reception at an antenna. The

phase center of an antenna is not constant but is dependent upon the observation angle and the signal frequency.

- **phase center**

The apparent center of signal reception at an antenna. The phase center of an antenna is not constant but is dependent upon the observation angle and the signal frequency.

- **phase difference processing**

relative positioning Computation of the relative difference in position between two points by the process of differencing simultaneous reconstructed carrier phase measurements at both sites. The technique allows cancellation of all errors which are common to both observers, such as clock errors, orbit errors, and propagation delays. This cancellation effect provides for determination of the relative position with much greater precision than that to which a single position (pseudorange solution) can be determined.

- **phase differencing**

The technique of using different GPS receivers at different locations to measure the phase angles of the carrier signal from the same satellite. These angles are compared by a communications link between the two locations if real-time operations are required.

- **phase interval**

In direct readout, the time between the end of a satellite image start tone and the start of the actual frame data. The phase interval represents white level video, interrupted by a black level pulse marking the start of each line and is used to set up phasing prior to image display.

- **phase lock**

The technique whereby the phase of an oscillator signal is made to follow exactly the phase of a reference signal by first comparing the phases of the two signals, and then using the resulting phase difference signal to adjust the reference oscillator frequency to eliminate phase difference when the two signals are next compared.

- **phase lock loop**

Another term for carrier-tracking loop; A type of electronic circuit used to demodulate satellite signals.

- **phase modulation**

Encoding information on a carrier signal by changing the phase so that some segments of the carrier are out of phase while others are in phase. With GPS, only two phase angles are used, 0 and 180, representing the two values, I or O.

- **phase-smoothed pseudo-range**
  The pseudo-range measurement which has had its "noise" level (random errors) reduced by being combined with the high precision carrier phase. It is still an unambiguous "range" measurement which can be processed using the standard algorithms of Point Positioning or Relative Positioning.

- **phenology**
  Subdiscipline of agriculture, a science that treats relations between climate and periodic biological phenomena that are related to or caused by climatic conditions, such as the budding of trees and the migration of birds.

- **photochemical smog**
  A type of smog that forms in large cities when chemical reactions take place in the presence of sunlight, its principal component is ozone. Ozone and other oxidants are not emitted into the air directly but form from reactions involving nitrogen oxides and hydrocarbons. Because of its smog-making ability, ozone in the lower atmosphere (troposphere) is often referred to as "bad" ozone.

- **photogeology**
  Mapping and interpretation of geologic features from aerial photographs.

- **photogrammetry**
  The science of making reliable measurements by the use of photographs and especially aerial photographs (as in surveying); Photogrammetry is an aerial remote sensing technique, the latest innovations of which employ a high-resolution aerial camera with forward motion compensation using global satellite navigation technology for pilot guidance over the designated photo block. Photogrammetry forms the baseline of many Geographic Information Systems (GIS) and Land Information System (LIS) studies and endeavors.

- **photograph**
  Representation of targets on film that results from the action of light on silver halide grains in the film's emulsion.

- **photographic ir**
  Short-wavelength portion (0.7 to 0.9 m) of the IR band that is detectable by IR color film or IR black-and-white film.

- **photographic uv**
  Long-wavelength portion of the UV band (0.3 to 0.4 m) that is transmitted through the atmosphere and is detectable by film.

- **photon**
  A quantum (smallest unit in which waves may be emitted or absorbed) of light.

- **photosynthetically active radiation**

Electromagnetic radiation in the part of the spectrum used by plants for photosynthesis.

- **physical climate system**

The system of processes that regulate climate, including atmospheric and ocean circulation, evaporation, and precipitation.

- **picture element**

In a digitized image, the area on the ground represented by each digital number. Commonly contracted to pixel.

- **pilotage**

The process of directing a vehicle by reference to recognizable landmarks or soundings, or to electronic or other aids to navigation.

- **pitch**

Rotation of an aircraft about the horizontal axis normal to its longitudinal axis that causes a nose-up or nose-down attitude.

- **pitch angle (pitch attitude)**

The angle between the longitudinal axis of the vehicle and the horizontal. See: roll.

- **pitch cue**

Flight director cue to control pitch; in fixed-wing aircraft, a yoke cue; in rotary-wing aircraft, a longitudinal cyclic cue.

- **pitch rate**

Rate of change of pitch; time derivative of pitch; Symbols: p; Symbols: theta dot; Typical Units: rad/s,deg/s; Dimensions: 1/Time;

- **pixel**

The smallest part (smallest addressable element) of an electronically-coded image, such as a computer display. Pixel is a contraction of "picture element."; A single display element on an LCD screen. The more pixels, the higher the resolution and definition; These are "picture elements," or little squares of color that come together to form shapes and images on a liquid crystal display (LCD). The more pixels per square inch, the sharper and more detailed picture you see.

- **planck's law**

An expression for the variation of emittance of a blackbody at a particular temperature as a function of wavelength.

- **planetary albedo**

The fraction of incident solar radiation thàt is reflected by a planet and returned to space. The planetary albedo of the Earth-atmosphere system is approximately 30 per cent, most of which is due to backscatter from clouds in the atmosphere.

- **planimetric map**

A map that shows only the horizontal positions of features

represented, as opposed to a topographic map.

- **plasma**

A fourth state of matter (in addition to solid, liquid, and gas) that exists in space. In this state, atoms are positively charged and share space with free negatively charged electrons. Plasma can conduct electricity and interact strongly with electric and magnetic fields. The solar wind is actually hot plasma blowing from the sun.

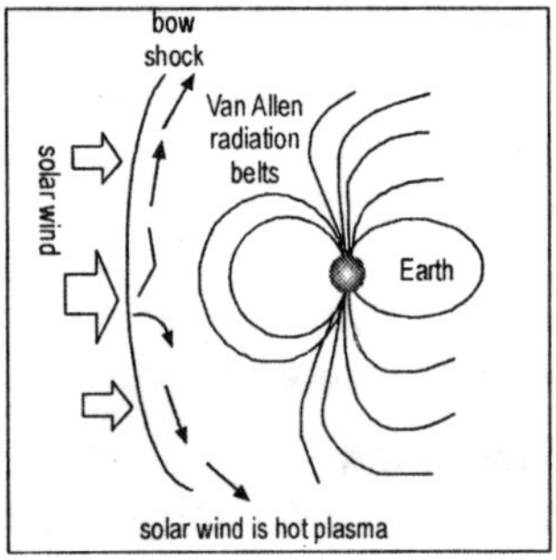

- **plat**

A map showing planned or actual features such as streets and building lots.

- **plate tectonics**

Concept that the Earth's crust is composed of rigid plates that move over a less rigid interior.

- **platform erection**

In the alignment of inertial systems, the process of bringing the vertical axis of a stable platform system into agreement with the local vertical.

- **platforms**

A satellite that can carry instruments. The same term is applied to automatic weather data transmitters installed on buoys, balloons, ships, and planes, and mounted in remote areas.

- **plotter display**

Provides an overhead "bird's-eye" view of your current position, relative to the waypoints and event marker/icons you've saved. A dotted-line marks the shortest route to the chosen waypoint, and a recorded plot-trail indicates the path you've taken so far.

- **plumb bob vertical**

The direction indicated by a simple, ideal, functionless pendulum that is motionless w.r.t. the earth; it indicates the direction of the vector sum of the gravitational and centrifugal accelerations of the earth at the location of the observer.

- **png-portable network graphics**

A file format for bitmapped graphic images approved by the World Wide Web Consortium as a replacement for GIF files. GIF files use a patented data compression algorithm; PNG is patent- and license-free.

- **pnt**

U.S. Space-Based Positioning, Navigation, and Timing Policy. A national policy that took effect on December 8, 2004 that establishes guidance and implementation actions for space-based positioning, navigation, and timing programs, augmentations, and activities for U.S. national and homeland security, civil, scientific, and commercial purposes.

- **pocket pc**

Software for personal digital assistants (PDAs) and any PDA that uses that software.

- **poes (polar-orbiting operational environmental satellite)**

Operated by the National Oceanic and Atmospheric Administration, they are designated "NOAA satellites." Included in this group are the current series of TIROS-N satellites, the third-generation polar-orbiting environmental spacecraft operated by NOAA.

- **point positioning**

Positions calculated by a single GPS receiver without reference to other receivers or known points. Position accuracy is typically 15 to 25 meters now that selective availability has been turned off; A geographic position produced from one receiver in a stand-alone mode. At best, position accuracy obtained from a stand-alone receiver is 15-25 meters, depending on the geometry of the satellites.

- **point spread function (psf)**

The image of a point source of radiation, such as a star, collected by an imaging device. A measure of the spatial fidelity of the device.

- **polar motion**

Motion of the instantaneous axis of the rotation of the Earth with respect to the solid body of the Earth. Irregular but more or less circular motion with an amplitude of about 15m and a main period of about 430 days (called Chandler Wobble).

- **polar mount**

Antenna mechanism permitting steering in both elevation and azimuth through rotation about a single axis. While an astronomer's polar mount has its axis parallel to that of the earth, satellite earth stations utilize a modified polar mount geometry that incorporates a declination offset.

- **polar orbit**

An orbit that passes close to the poles, thereby enabling a satellite to pass over most of the surface, except the immediate vicinity of the poles themselves.

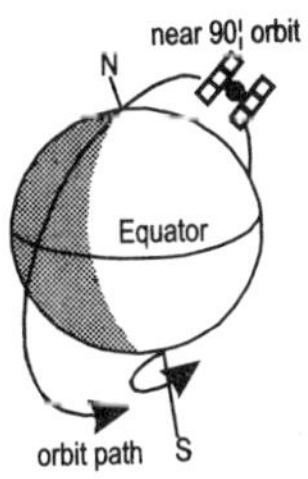

An orbit with its plane aligned in parallel with the polar axis of the earth; An orbit with an orbital inclination of near 90 degrees where the satellite ground track will cross both polar regions once during each orbit. The term is used to describe the near-polar orbits of spacecraft such as the USA's NOAA/TIROS and Landsat satellites.

- **polarization**

The direction of orientation in which the electrical field vector of electromagnetic radiation vibrates; A technique used by the satellite designer to increase the capacity of the satellite transmission channels by reusing the satellite transponder frequencies. In linear cross polarization schemes, half of the transponders beam their signals to earth in a vertically polarized mode; the other half horizontally polarize their down links. Although the two sets of frequencies overlap, they are 90 degree out of phase, and will not interfere with each other. To successfully receive and decode these signals on earth, the earth station must be outfitted with a properly polarized feedhorn to select the vertically or horizontally polarized signals as desired. In some installations, the feedhorn has the capability of receiving the vertical and horizontal transponder signals simultaneously, and routing them into separate LNAs for delivery to two or more satellite television receivers. Unlike most domestic satellites, the Intelsat series use a technique known as left-hand and right-hand circular polarization.

- **polarization rotator**

A device that can be manually or automatically adjusted to select one of two orthogonal polarizations.

- **polarized radiation**

Electromagnetic radiation in which the electrical field vector is contained in a single plane, instead of having random orientation relative to the propagation vector: Most commonly refers to radar images.

- **poor gdop**

Poor GDOP, a large value representing a small unit vector-volume, results when angles from receiver to the set of SVs used are similar.

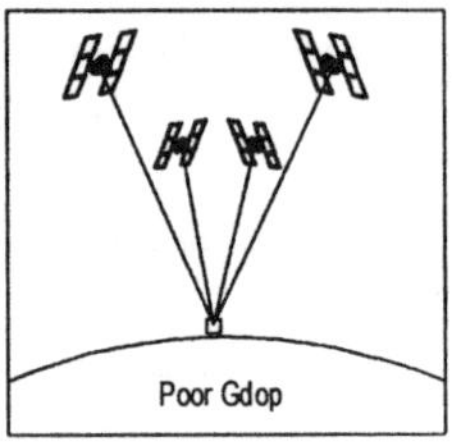

- **pop (point-of-presence)**

In the field of telephony, POP refers to the facility interface between the Local Exchange

Carrier (LEC) and the Inter-Exchange Carrier (IXC). A POP must be established on each trunk path that connects switches that belong to the local and long-distance carriers. On one side of the POP, the LEC is responsible for service; on the other side, the IXC is responsible.

- **position**
  Location, usually in fixed earth coordinates such latitude and longitude; location, either scalar or vector, often with subscripts such as ENU or XYZ to denote source or coordinate frame; time integral of velocity; Symbols: p, P, x, y, z; Typical Units: ft,nmi; Dimensions: Length; The 3-D coordinates of a point, usually given in the form of Latitude, Longitude, and Altitude (or Ellipsoidal Height), though it may be provided in the 3-D Cartesian form, or any other transformed map or geodetic reference system. An estimate of error is often associated with a position.

- **position dilution of precision (pdop)**
  Measure of the geometrical strength of the GPS satellite configuration for 3-D positioning.

- **position display**
  One of the primary navigational data screens that emphasizes the present position latitude/ longitude coordinates, as well as other helpful information.

- **position error**
  The distance that the displayed position is off by.

- **position fix**
  The GPS receiver's computed position coordinates.

- **position fixing**
  Using a coordinate system to determine a point on a map's location.

- **position format**
  The way in which the GPS receiver's position will be displayed on the screen. Commonly displayed as latitude/ longitude in degrees and minutes, with options for degrees, minutes and seconds, degrees only, or one of several grid formats.

- **position update**
  To cause navigation sensors, devices, or algorithms to reset position to value known to be more accurate due to inaccuracies and drift in the devices and algorithms

- **positional dilution of precision**
  Measure of the geometrical strength of the GPS satellite configuration; A unit less figure of merit expressing the relationship between the error in user position and the error in satellite position, which is a function of the configuration of satellites from which signals are derived in positioning. Geometrically, PDOP is proportional to 1 divided by the

volume of the pyramid formed by lines running from the receiver to four observed satellites. Small values, such as "3", are good for positioning while higher values produce less accurate position solutions. Small PDOP is associated with widely separated satellites

- **positive photograph**
  Photographic image in which the tomes are directly proportional to the terrain brightness.

- **post processed differential gps**
  Process of differentially correcting GPS data after it has been collected by differential correction software; In post-processed differential GPS the base and roving receivers have no active data link between them. Instead, each records the satellite observations that will allow differential correction at a later time. Differential correction software is used to combine and process the data collected from these receivers.

- **post processing**
  Post processing is the reduction, calculation and processing of global satellite data after the data has been collected in the field. Post processing is usually accomplished on a computer in an office environment where appropriate software is employed to achieve optimum position solutions.

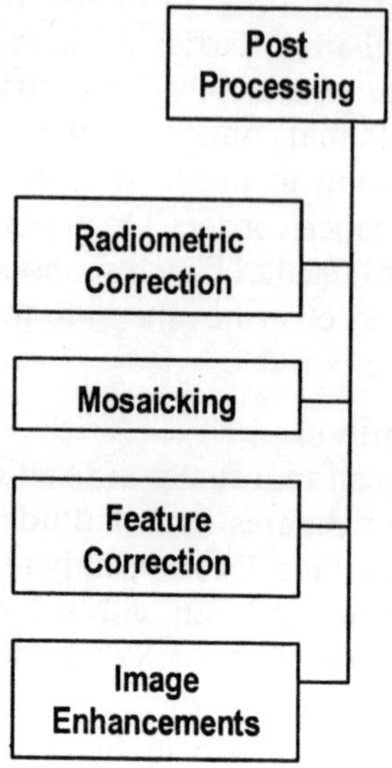

- **precession**
  The comparatively slow torquing of the orbital planes of all satellites with respect to the Earth's axis, due to the bulge of the Earth at the equator which distorts the Earth's gravitational field. Precession is manifest by the slow rotation of the line of nodes of the orbit (westward for inclinations less than 90 degrees and eastward for inclinations greater than 90 degrees).

- **precipitation**
  Moisture that falls from clouds. Although clouds appear to float in the sky, they are always falling, their water droplets slowly being pulled down by gravity. Because their water droplets are so small and light, it can take 21 days to fall 1,000 feet and wind currents can

easily interrupt their descent. Liquid water falls as rain or drizzle. All raindrops form around particles of salt or dust. Water or ice droplets stick to these particles, then the drops attract more water and continue getting bigger until they are large enough to fall out of the cloud. Drizzle drops are smaller than raindrops. In many clouds, raindrops actually begin as tiny ice crystals that form when part or all of a cloud is below freezing. As the ice crystals fall inside the cloud, they may collide with water droplets that freeze onto them. The ice crystals continue to grow larger, until large enough to fall from the cloud. They pass through warm air, melt, and fall as raindrops. When ice crystals move within a very cold cloud (10 degrees F and -40 degrees F) and enough water droplets freeze onto the ice crystals, snow will fall from the cloud. If the surface temperature is colder than 32 degrees F, the flakes will land as snow. Precipitation Weights:

one raindrop .000008 lbs

one snowflake .0000003 lbs

one cumulus cloud 10,000,000 lbs

one thunderstorm 10,000,000,000 lbs

one hurricane 10,000,000,000,000 lbs

- **precise positioning service (pps)**

The GPS positioning, velocity, and time service which is available on a continuous, worldwide basis to users authorized by the U.S. Department of Defence (typically using P-Code); The most accurate Absolute Positioning possible with GPS navigation receivers, based on the dual-frequency encrypted P-Code. Available to the military users of GPS. Typical accuracy is of the order of 10-20m.

- **precision**

Measure of exactness, possibly expressed in number of digits, for example, computed to the nearest millimeter; The degree of repeatability that repeated measurements of the same quantity display, and is therefore a means of describing the quality of the data with respect to random errors. Precision is traditionally measured using the standard deviation.

- **precision (p)-code**

A PRN code transmitted by GPS satellites. The code consists of about 2.35 3 1014 chips and is sent at a rate of 10.23 megabits per second. At this rate, it would take 266 days to transmit the complete code. Each satellite is assigned a unique one-week segment of the code that is reset at Saturday/ Sunday midnight. The P-code is currently transmitted on both the L1 and L2 frequencies; A very long sequence of pseudo-random binary biphase modulations on the GPS carrier at a chip rate of

10.23 MHz, which repeats about every 267 days. Each 1-week segment of this code is unique to one GPS satellite and is reset each week

- **prescribed interexchange charge (picc)**
  The charge the local exchange company assesses the long distance company when a consumer picks it as his or her long distance carrier.

- **present position**
  Current location on the face of the Earth, in terms of the specific latitude/longitude coordinates, displayed in degrees/minutes/ & thousandths of a minute.

- **present status words**
  Most recently reported status words.

- **prevailing westerlies**
  Winds in the middle latitudes (approximately 30 degrees to 60 degrees) that generally blow from west to east. The subtropical high pressure regions at the horse latitudes (30 degrees) forces surface air poleward, and the rotation of the Earth causes these winds to bear to the right (east) in the Northern Hemisphere and to the left (east) in the Southern Hemisphere. This is, to some extent, an idealized picture of the atmospheric circulation. The actual circulation on individual days includes modifications and variations due to the migratory cyclones and anticyclones of middle latitudes, causing rapid and often violent weather changes, as warm semi-tropical air from the horse latitudes meets cold polar air from the high latitudes.

- **previsual symptom**
  A vegetation anomaly that is recognizable on IR film before it is visible to the naked eye or on normal color photographs. It results when stressed vegetation loses its ability to reflect photographic IR energy and its recognizabie on IR color film by a decrease in brightness of the red hues.

- **primary colors**
  A set of three colors that,in various combinations will produce the full range of colors in the visible spectrum. There are two sets of primary colors, additive and subtractive.

- **primary flight control system (pfcs)**
  The most basic part of the flight controls operated by a pilot, including wheel (fixed wing), yoke (fixed wing), cyclic (rotary wing), pedals (fixed wing and rotary wing), throttle (fixed wing), and collective (rotary wing);

- **primary units**
  A standard set of four units to which all units can be resolved; primary units are Mass(M),

Length(L), Time(theta), and Temperature(T); for example, standard units for velocity might be kt (nmi/hr), ft/s, m/s, mph, but primary are always Length / Theta; sometimes written with negative subscripts.

- **prime meridian**
  An imaginary line running from north to south through Greenwich, England, used as the reference point for longitude; The zero meridian, used as a reference line from which longitude east and west is measured. It passes through Greenwich, England; The reference line for Longitude in the Lat/Long grid system. All other meridians of longitude are designated in degrees East or West of the Prime Meridian up to 189?. The Prime Meridian is 0? Longitude.

- **prime radius of curvature**
  Radius of the earth in the east/ west direction at a given position; Synonyms: earth radius east/ west; Symbols: rho sub P; Typical Units: ft; Dimensions: Length;

- **principal component analysis**
  The analysis of covariance in a multiple data set so that the data can be projected as additive combinations on to new axes, which express different kinds of correlation among the data.

- **principal point**
  Optical center of an aerial photograph.

- **principal-component (pc) image**
  Digitally processed image produced by a transformation that recognizes maximum variance in multispectral images.

- **printout**
  Display of computer data in alphanumeric format.

- **prismatic compass**
  A compass that allows the user to see the compass face and distant locations at the same time.

- **prn**
  Pseudo-random noise, a sequence of digital 1's and 0's which appears to be randomly distributed like noise, but can be exactly reproduced. Each NAVSTAR satellite has its own unique C/A and P pseudo-random-noise codes and are often referred to by their PRN number.

- **prn number**
  A number assigned by the GPS system designers to a given set of pseudorandom codes. Typically, a particular satellite will keep its PRN (and hence its code assignment) indefinitely, or at least for a long period of time. It is commonly used as a way to label a particular satellite.

- **prn-pseudo-random noise**
  A digital signal which appears to be randomly distributed like noise, but is actually not random. Each NAVSTAR satellite has its own code.

- **probability density function (pdf)**
  A function indicating the relative frequency with which any measurement may be expected to occur. In remote sensing it is represented by the histogram of DN in one band for a scene.

- **process**
  An association of phenomena governed by physical, chemical, or biological laws. An example of a process is the vertical mixing of ocean waters in the so-called surface-mixed layer; the state variables for this process include temperature, salinity in the water on a vertical scale of tens of meters, and heat flow and wind stress at the sea surface. Other examples include the volcanic deposition of dust and gases into the atmosphere, eddy formation in the atmosphere and oceans, and soil development.

- **process study**
  An organized, systematic investigation of a particular process designed to identify all of the state variables involved and to establish the relationships among them. Process studies yield numerical algorithms that connect the state variables and determine their rates of change; such algorithms are essential ingredients of Earth system models.

- **prograde orbit**
  Orbits of the Earth in the same direction as the rotation of the Earth (west-to-east).

- **programmable windows**
  The ability to customize existing split-panel window groupings with the specific combination of navigational data preferred.

- **progressive scanning**
  The method for tracing scan lines onto a computer monitor whereby all scanlines are presented in sequence 60 or more times per second. Unlike the NTSC video standard, there is no single standard for computer display. Most computer monitors will accept a wide range of screen resolutions and scanning rates utilizing multiscan technologies.

- **protocol**
  Any set of standard procedures that permit devices to inter-communicate.

- **pseudo random code**
  A signal with random noise-like properties. It is a very complicated but repeating pattern of 1's and O's.

- **pseudo-kinematic surveying**
  A variation of the kinematic method where roughly five-minute site occupations are repeated at a minimum of once each hour.

- **pseudolite**
  An Earth-based transmitter designed to mimic a satellite. May be used to transmit differential corrections; A ground-based differential GPS receiver which transmits a signal like that of an actual GPS satellite, and can be used for ranging. Originally intended as an augmentation for Local Area Augmentation Systems to aid aircraft landings. However, pseudolites may also be used where signal obstructions are such that insufficient GPS satellites can be tracked. In fact, pseudolites are feasible in circumstances where no satellite signals are observable, e.g. for indoor applications.

- **pseudolite-pseudo-satellite**
  A ground-based differential GPS receiver that simulates the signal of a GPS satellite and can be used for ranging. The data portion of the signal may also contain differential corrections that can be used by receivers to correct for GPS errors.

- **pseudorandom code**
  The identifying signature signal transmitted by each GPS satellite and mirrored by the GPS receiver in order to separate and retrieve the signal from background noise.

- **pseudo-random noise (prn)**
  A signal with random noise-like properties. It is very complicated but repeated pattern of 1's and 0's; A binary signal with random noise-like properties. It is generated by mathematical algorithm or "code", and consists of repeated pattern of 1's and 0's. This binary code can be modulated on the GPS carrier waves using Binary Shift-Key (BSK) modulation. The C/A-Code and the P-Code are examples of PRN codes. Each satellite transmits a unique C/A-Code and P-Code sequence (on the same L1 and L2 frequencies), and hence a satellite may be identified according to its "PRN number", e.g. PRN2 or PRN14 are particular GPS satellites.

- **pseudorange**
  The calculated range from the GPS receiver to the satellite determined by taking the difference between the measured satellite transmit time and the receiver time of measurement, and multiplying by the speed of light. Contains several sources of error; The measured distance between the GPS receiver and the GPS satellite using uncorrected time comparisons from satellite-transmitted code and the local receiver's reference code; A measure of the apparent propagation time from the satellite to the receiver antenna, expressed as a distance. The apparent propagation time is determined from the time shift required to align a replica of the

GPS code generated in the receiver with the received GPS code. The time shift is the difference between the time of signal reception (measured in the receiver time frame) and the time of emission measured in the satellite time frame). Pseudorange is obtained by multiplying the apparent signal-propagation time by the speed of light. Pseudorange differs from the actual range by the amount that the satellite and receiver clocks are offset, by propagation delays, and other errors including those introduced by selective availability; A measure of the apparent propagation time from the satellite to the receiver antenna, expressed as a distance. Pseudorange is obtained by multiplying the apparent signal-propagation time by the speed of light. Pseudorange differs from the actual range by the amount that the satellite and user clocks are offset, by propagation delays, and other errors. The apparent propagation time is determined from the time shift required to align (correlate) a replica of the GPS code generated in the receiver with the received GPS code. The time shift is the difference between the time of signal reception (measured in the receiver time frame) and the time of emission (measured in the satellite time frame).

- **pseudorange difference**
  See reconstructed carrier phase.

- **pseudorange measurements**
  Measurements made using one of the pseudorandom codes on the GPS signals. They provide an unambiguous measure of the range to the satellite including the effect of the satellite and user clock biases.

- **pseudo-static gps**
  Also known as pseudo-kinematic and repeat occupation, this relative positioning technique relies upon two or more simultaneous observations at a point pair, separated by some time interval (typically 60 minutes or more), in order to solve the integer bias terms from the change in satellite geometry occurring between the repeat observations. GPS has become an important tool for high accuracy relative positioning since it was introduced in the early 1980s. In order to ensure high accuracy, both the functional model and the stochastic model must be correctly defined. Since GPS measurements are contaminated by many errors, it is impossible to model all systematic errors in the functional model without some understanding, or prior knowledge, of the physical phenomena which underpin these errors.

- **psychrometer**
  An instrument designed to measure dew point and relative

humidity, consisting of two thermometers (one dry bulb and one wet bulb). The dew point and humidity levels are determined by drying the wet bulb (either by fanning or whirling the instrument) and comparing the difference between the wet and dry bulbs with preexisting calculations.

- **ptt - post telephone and telegraph administration**

  Refers to operating agencies directly or indirectly controlled by governments in charge of telecommunications services in most countries of the world.

- **public switched telephone network (pstn)**

  Any common carrier network that provides circuit switching among public users. Note: The term is usually applied to the public switched telephone network, but it could be applied more generally to other switched networks, e.g., packet switched public data networks.

- **pulse**

  Short burst of electromagnetic radiation transmitted by a radar antenna.

- **pulse code modulation**

  A time division modulation technique in which analog signals are sampled and quantized at periodic intervals into digital signals. The values observed are typically represented by a coded arrangement of 8 bits of which one may be for parity.

- **pulse length**

  Duration of a burst of energy transmitted by a radar antenna, measured in microseconds; The length of time a sonar unit transmits a pulse of sound into the water. Sometimes referred to as Pulse Width.

- **pushbroom scanner**

  An alternate term for an along-track scanner

- **pushbroom system**

  An imaging device consisting of a fixed linear array of many sensors which is swept across an area by the motion of the platform, thereby building up an image. It relies on sensors whose response and reading is nearly instantaneous, so that the image swathe can be segmented into pixels representing small dimensions on the ground.

- **qpsk (quadrature phase shift keying)**

  Used to modulate digital information onto an RF carrier when satellite transponders are used to transmit MPEG 2 signals. Rather than using the amplitude or frequency of the carrier to convey the information, QPSK modulates the phase of the carrier signal. Depending on the data being modulated, the carrier is forced into one of four different phase states known as a symbol. The advantage of this method is that each symbol contains two data bits, thus doubling the potential amount of data that is transmitted over conventional amplitude or frequency modulation techniques.

- **quadrifilar helix**
  A type of circularly polarized antenna used with some GPS receivers. The antenna consists of two orthogonal, fractional-turn bifilar wire or metal strip helices fed in phase quadrature

- **quadrifilar helix antenna**
  A type of GPS antenna in which four spiraling elements form the receiving surface of the antenna. For GPS use, quadrifilar antennas are typically half-wavelength or quarter-wavelength size and encased in a plastic cylinder for durability.

- **quantum**
  The elementary quantity of EM energy that is transmitted by a particular wavelength. According to the quantum theory, EM radiation is emitted, transmitted, and absorbed as numbers of quanta, the energy of each quantum being a simple function of the frequency of the radiation.

- **quartz oscillator**
  Quartz oscillator is the timing device within a receiver that synchronizes the operation and maintains time for the receiver.

- **quaternion**
  A system of representing attitude by measuring angle of aircraft center line with respect to three orthoginal axes plus rotation about centerline; quaternions are used over Euler angles (pitch, roll, yaw) when pitch can approach 90° because of a singularity on Euler angles at 90deg; discrete-time computations using quaternions can run more slowly than those with Euler angles while producing results of the same accuracy.

- **r95**
  A position accuracy measure. The R95 value is defined as a circle's radius, when centred at the true position, encloses 95% of the data points in a horizontal scatter plot.

- **radar**
  Acronym for radio detection and ranging. Radar is an active form of remote sensing that operates in the microwave and radio wavelength regions.

- **radar altimeter**
  A non-imaging device that records the time of radar returns from vertically beneath a platform to estimate the distance to and hence the elevation of the surface; carried by Seasat and the EAS-ERS-1 platforms.

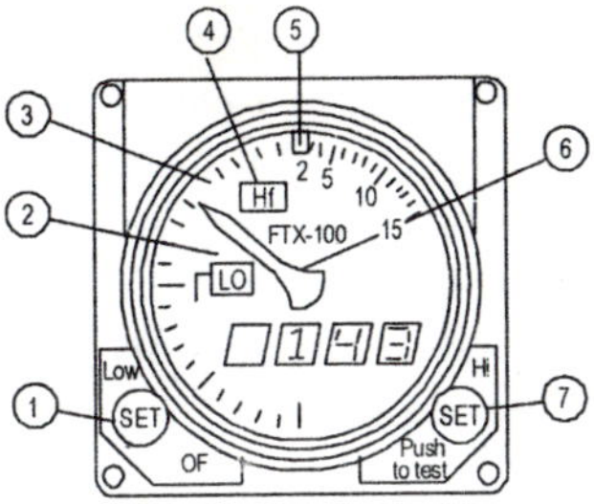

- **radar altitude**
  Height with respect to the terrain below (distance above closest dirt); Synonyms: above ground level; Symbols: h sub r; Typical Units: ft; Dimensions: Length.

- **radar altitude select (ralt sel)**
  A basic guidance mode, providing vertical guidance to an operator selected radar altitude;

- **radar cross section**
  A measure of the intensity of backscattered radar energy from a point target. Expressed as the area of a hypothetica surface which scatters radar equally in all directions and which would return the same energy to the antenna.

- **radar scattering coefficient**
  A measure of the back-scattered energy from a target with a large area. Expressed as the average radar cross section per unit area in decibels (dB). It is the fundamental measure of the radar properties of a surface.

- **radar scatterometer**
  A non-imaging device that records radar energy backscattered from terrain as a function of depression angle.

- **radar shadow**
  Dark signature on a radar image representing no signal return. A shadow extends in the far-range direction form an object that intercepts the radar beam.

- **radial error probability (rep)**
  A probability that a percentage of one-dimension measurements will lie on a radial (line) of given length, with the origin centered at truth or mean of the measurements; used to specify test cases for measurement errors of sensors of one dimension, such as vertical velocity.

- **radial relief displacement**
  The tendency of vertical objects to appear to learn radially away from the center of a vertical aerial photograph. Caused by the conical field of view of the camera lens.

- **radian**
  Nagle subtended by an arc of a circle equal in length to the radius of the circle 1 rad = 57.3i.

- **radiance**
  Radiance is the power of EM radiation incident on unit area per unit solid angle in a given direction.

- **radiant**
  1. In optics, the point or object from which light proceeds.
  2. In geometry, a straight line proceeding from a given point,

or fixed pole, about which it is conceived to revolve.

3. In astronomy, the point in the heavens from which a shower of meteors seems to proceed.

- **radiant energy peak**

Wavelength at which the maximum electromagnetic energy is radiated at a particular temperature.

- **radiant flux**

Rate of flow of electromagnetic radiation measured in watts per square centimeter.

- **radiant temperature**

Concentration of the radiant flux from a material. Radiant temperature is the kinetic temperature multiplied by the emissivity to the one-fourth power.

- **radiation**

Propagation of energy in the form of electromagnetic waves; Energy transfer in the form of electromagnetic waves or particles that release energy when absorbed by an object.

- **radiation budget**

A measure of all the inputs and outputs of radiative energy relative to a system, such as Earth.

- **radiative cooling**

Cooling process of the Earth's surface and adjacent air, which occurs when infrared (heat) energy radiates from the surface of the Earth upward through the atmosphere into space. Air near the surface transfers its thermal energy to the nearby ground through conduction, so that radiative cooling lowers the temperature of both the surface and the lowest part of the atmosphere.

- **radiative transfer**

Theory dealing with the propagation of electromagnetic radiation through a medium.

- **radio direction finder**

A radio receiver that features a directional antenna and a visual null indicator for use in determining lines of position from radiobeacons at known positions

- **radio frequency**

Radio frequency refers to a signal generated by a radio transmitter and sent out through an antenna.

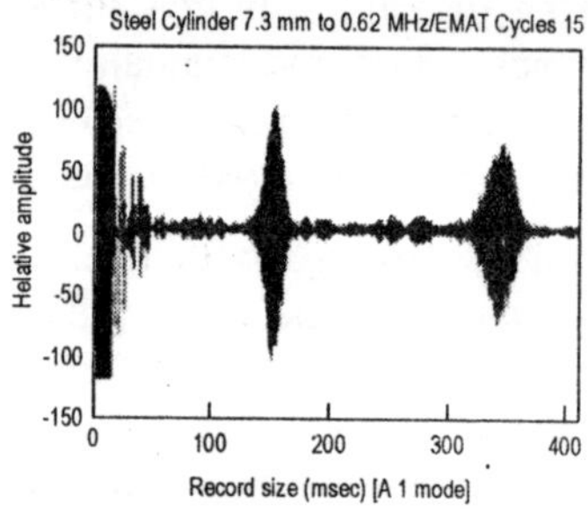

The frequency of the transmission is described in terms of the number of cycles per second or Hertz (Hz). A radio would be tuned to this frequency in order to receive the transmission. A radio signal is sometimes referred to by its initials, "RF"

- **radio frequency interference (rfi)**

A term used to denote apparent disturbance or distortion to satellite signals.

- **radio navigation**

Navigation relative to radio station, providing, for example, of relative bearing, range, lateral deviation, and glideslope; Examples include VOR, TACAN, and PLS. Radio navigation differs from other navigation in that the transmitter signals often dropout for a long period of time, like minutes. This can occur because of natural obstructions, or because the transmitter was shut down intentionally. In hostile territory, a PLS can locate a downed pilot, who would be foolhardy to be continuously transmitting, but would transmit infrequently with small bursts of data. The Radio-Navigation system accommodates this phenomenon by simulating range and bearing to the fixed site when it is not transmitting. After reacquiring a mobile transmitter, the mobile station's position is re-determined. The navigation component supports wash-out filters on output data.

- **radio spectrum**

The complete range of frequencies or wave lengths of electromagnetic waves, specifically those used in radio and television.

1. AM Radio = 535kHz - 1605kHz (MF)
2. TV Band I (Channels 2 - 6) = 54MHz - 88MHz (VHF)
3. FM Radio Band II = 88MHz - 4. 108MHz (VHF)
4. TV Band III (Channels 7 - 13) = 174MHz - 216MHz (VHF)
5. TV Bands IV & V (Channels 14 - 69) = 470MHz - 806MHz (UHF)

- **radio technical commission for maritime services (rtcm) special committee 104**

A committee created for the purposes of establishing standards and guidance for interfacing between radio beacon-based data links and GPS receivers, and to provide standards for ground-based differential GPS stations.

- **radio technical committee for maritime applications (rtcm)**

RTCM Special Committee 104 has developed standard message types for use by differential GPS transmitting stations. The message content has been defined and hence when the RTCM-104 standard (version 2.2 is the latest) is implemented within a user receiver, it is able to decode and apply the DGPS corrections to its raw data in order to generate a DGPS-corrected coordinate.

- **radio wave**

An electrical impulse sent through the atmosphere at radio frequency.

- **radioactive**

Giving off or capable of giving off radiant energy in the form of particles or rays, as in alpha, beta, and gamma rays.

- **radiometer**

Device for quantitatively measuring radiant energy, especially thermal radiation; An instrument that quantitatively measures electromagnetic radiation. Weather satellites carry radiometers to measure radiation from snow, ice, clouds, bodies of water, the Earth's surface, and the sun.

- **radiosonde**

A balloon-borne instrument that measures meteorological parameters from the Earth's surface up to 20 miles in the atmosphere. The radiosonde measures temperature, pressure, and humidity, and transmits or "radios" these data back to Earth. Upper air winds also are determined through tracking of the balloon ascent. Radiosonde observations generally are taken twice a day (0000 and 1200 UTC) around the globe. NOAA's National Weather Service (NWS) operates a network of about 90 radiosonde observing sites in the U.S. and its territories. When the balloons burst, radiosondes return to Earth on a parachute. Approximately 25 per cent are recovered and returned to NWS for reconditioning and reuse.

- **raim**

Receiver Autonomous Integrity Monitoring; A GPS receiver system that would allow the receiver to detect incorrect signals being transmitted by the satellites by comparing solutions with different sets of satellites; Technology added to aviation GPS used to detect questionable data from one or more satellites. It works by over-determining position using at least five satellites or four satellites and a barometric altitude input.

- **rain forest**

An evergreen woodland of the tropics distinguished by a continuous leaf canopy and an average rainfall of about 100 inches per year. Rain forests play an important role in the global environment. The Earth sustains life because of critical balances and interactions among many factors. Were there not processes at work that limit the effects of other essential processes, Earth would become uninhabitable. Destruction of tropical rain forests reduces the amount of leaf area in the tropics, and consequently the amount of carbon dioxide absorbed, causing increases in levels of carbon dioxide and other atmospheric gases. It is estimated that cutting and burning of tropical forests contributes about 20 per cent of the carbon dioxide added to the atmosphere each year.

- **rain gauge**

  Calibrated container that measures the amount of rainfall during a specific period of time.

- **rain outage**

  Loss of signal at Ku or Ka Band frequencies due to absorption and increased sky-noise temperature caused by heavy rainfall.

- **ram (random access memory)**

  In computers, the main system memory, usually consisting of volatile memory (memory that loses its data when power is removed) solid-state chips; Computers use two types of memory, RAM and ROM. RAM is the computer's working area, the primary location where the microprocessor stores the information it needs. The designation "random access" stems from the microprocessor's ability to access information in memory randomly by knowing its location, or address, rather than hunting through memory sequentially from beginning to end. Because information in RAM is stored electronically, accessing data stored in RAM is much faster than getting that data from a mechanical storage device such as a disk drive. But because it is stored electronically, all information in RAM is temporary (which is why you must store it on a more permanent storage capability, such as a disk). Compare with ROM.

- **random line dropout**

  In scanner images, the loss of data from individual scan lines in a nonsystematic fashion.

- **range**

  In radar usage this is the distance in the direction of radar propagation, usually to the side of the platform in an imaging radar system. The slant range is the direct distance from the antenna to the object, whereas the distance from the ground track of the platform to the object is termed the ground range.

- **range rate**

  The rate of change of range between the satellite and receiver. The range to a satellite changes due to satellite and observer motions. Range rate is determined by measuring the Doppler shift of the satellite beacon carrier.

- **range resolution**

  In radar images, the spatial resolution in the range direction, which is determined by the pulse length of the transmitted microwave energy.

- **ranging**

  A technique used to determine a line of position by calculating the distance between a receiver and a known reference point.

- **raster**

  The scanned and illuminated area of a video display, produced by a modulated beam of electrons sweeping the phosphorescent screen line by line from top to bottom at a regular rate of repetition; Computer graphics term describing a predetermined pattern of line that provides uniform coverage of a display space.

- **raster format**

  A means of representing spatial data in the form of a grid of DN, each line of which can be used to modulate the lines of a video raster.

- **raster maps**

  Raster maps are commonly referred to as "bit mapped" and are in a GIF or JPEG format. Raster images are well-suited for web sites where editing or resizing are not required. They can be made smaller, but are not meant to be enlarged or customized in the ways that vector maps can be manipulated.

- **raster pattern**

  Pattern of horizontal lines swept by an electron beam across the face of a CRT that constitute the image display.

- **rate limiter**

  A filter that passes the input as the output, except that rate of change of the output is limited to a maximum absolute value.

- **ratio image**

  An image prepared by processing digital multi-spectral data as follows: for each pixel, the value for one band is divided by that of another. The resulting digital values are displayed as an image.

- **raw**

  Data taken directly from the sensor; Compare: derived, estimated, filtered, measured, selected, smoothed

- **rayleigh criterion**

  In radar, the relationship between surface roughness, depression angle, and wavelength that determines whether a surface will respond in a rough or smooth fashion to the radar pulse.

- **rayleigh scattering**

  Selective scattering of light in the atmosphere by particle that are small compared with the wavelength of light.

- **rboc (regional bell operating company)**

  An independent common carrier that provides local exchange services within specific areas. The term originally referred to the local telephone service companies which were created as a result of the break-up of AT&T in 1984. The term now refers to all local exchange services.

- **rds (radio data system)**
Digital channel transmitted within the FM radio wavelength used to relay traffic messages.

- **rds-tmc**
Radio Data Systems incorporating a Traffic Message Channel.

- **real time**
Time in a computational process which runs at the same rate as a physical process; for example, algorithms designed to run a fixed period t (filter time constants at set for t) and actually execute with frequency 1/t execute in real time; Avionics systems must run in real time; Refers to images or data made available for inspection simultaneously with their acquisition.

- **real time kinematic (rtk)**
The DGPS procedure whereby carrier phase corrections are transmitted in real time from a reference station to the user's roving receiver; The Relative Positioning procedure whereby carrier phase measurements (or corrections) are transmitted in real-time from a Reference or Base Station to the user's roving receiver. Centimetre accuracy is achieved without the need to record and post-process double-differenced carrier phase observables.

- **real-aperture radar**
Radar system in which azimuth resolution is determined by the transmitted beam width, which is in turn determined by the physical length of the antenna and by the wavelength.

- **real-time dgps**
A Base Station computes, formats, and transmits pseudo-range corrections via some sort of data communication link (e.g., VHF or UHF radio, cellular telephone, FM radio sub-carrier or satellite com link). The roving receiver requires some sort of data link receiving equipment to receive the transmitted DGPS corrections so that they can be applied to its current observations. Several aircraft manufacturers are developing aircraft flight test reference systems often called TSPI systems based upon DGPS technology. Requests have been received, and additional requests are expected, for approval of these systems for use during noise certification tests. Typically, the hardware components of these systems are: GPS receivers and antennae on the ground and in the aircraft; data link transmitter and antenna on the ground and corresponding receiver and antenna in the aircraft; a laptop computer in the aircraft; and batteries and electronic power supplies.

- **real-time engineering simulation (rtes)**

  A simulator designed to test avionics algorithms with a pilot in the loop, consisting of a simulated cockpit, an aircraft model, sensor models, and algorithms to be tested; rtes is often used during development to check algorithms, such as the navigation, radio navigation, guidance and flight director, prior to full-scale software development; rtes gives systems engineers early insight in human factors problems, pilot complaints, algorithm bugs, unstability in the algorithms, expected performance (accuracy), and pilot-machine interaction.

- **real-time ztm**

  Thales Navigation's unique technology that includes Carrier Phase Differential (CPD) processing to provide On The Fly (OTF) ranging data acquisition and differential processing in its professional GPS and GNSS solutions.

- **reasonableness**

  A test to determine if data is reasonable, for example, radar altitude must be positive, and two devices should return similar data within known limits of each other.

- **receiver (rx)**

  An electronic device which enables a particular satellite signal to be separated from all others being received by an earth station, and converts the signal format into a format for video, voice or data.

- **receiver autonomous integrity monitoring**

  A form of receiver self-checking in which redundant pseudo-range observations are used to detect if there is a problem or "failure" with any of the measurements -- only four measurements are needed to derive 3-D coordinates and the receiver clock error, hence any extra measurements can be used for checking.

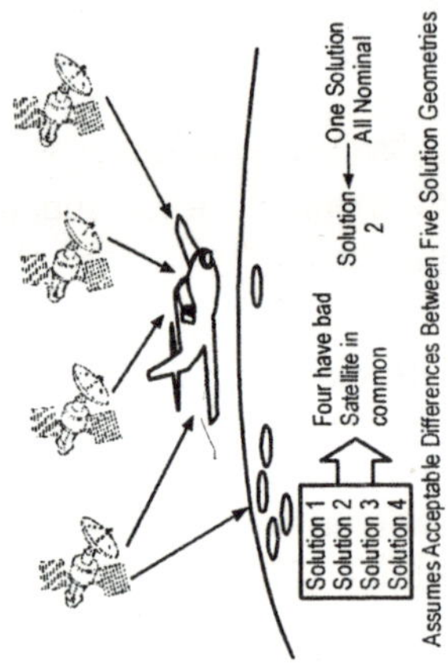

Once the failed measurements have been identified they may be eliminated from the navigation fix. RAIM is a concept that has been introduced by aviation users who are concerned that GPS does not have the level of Integrity necessary for non-precision airport approaches or GPS-aided landing: A technique by which a GPS receiver system detects incorrect satellite signals by comparing solutions with different sets of satellites.

- **receiver channels**
  A GPS receiver specification which indicates the number of independent hardware signal processing channels included in the receiver design.

- **receiver sensitivity**
  The ability of a receiver to detect weak signals through the noise level of the receiving system, which includes the antenna and internal thermal noise of the receiver; Expressed in dBm this tells how much power the detector must receive to achieve a specific baseband performance, such as a specified bit error rate or signal to noise ratio.

- **receiver-independent exchange (rinex)**
  A set of standard definitions and formats to promote the free exchange of GPS data and facilities the use of data from any GPS receiver with any software package. The format includes definitions for time, phase, and range.

- **recognizability**
  Ability to identify an object on an image.

- **reconnaissance**
  A preliminary survey to gain information (from Middle French - literally, recognition).

- **reconstructed carrier phase**
  The difference between the phase of the incoming Doppler-shifted GPS carrier and the phase of a nominally constant reference frequency generated in the receiver. For static positioning, the reconstructed carrier phase is sampled at epochs determined by a clock in the receiver. The reconstructed carrier phase changes according to the continuously integrated Doppler shift of the incoming signal, biased by the integral of the frequency offset between the satellite and receiver reference oscillators. The reconstructed carrier phase can be related to the satellite-to-receiver range, once the initial range (or phase ambiguity) has been determined. A change in the satellite-to-receiver range of one wavelength of the GPS carrier (19 cm for L1) will result in a one-cycle change in the phase of the reconstructed carrier.

- **rectilinear**
  Refers to images with no geometric distortion in which the scales in the horizontal and vertical directions are identical.

- **redundancy**
  Information on an image which is either not required for interpretation or cannot be seen. Redundancy may be spatial or spectral. The term also refers to multispectral data where the

degree of correlation between bands is so high that one band contains virtually the same information as all the bands.

- **reference direction**

  A direction from which other directions are reckoned; for example, true north, grid north, etc.

- **reference line**

  A line from which angular or linear measurements are reckoned.

- **reference network**

  A reference network is a series of monuments or reference points with accurately measured mutual vectors/distances used as a reference basis for cadastral and other types of survey.

- **reference satellite**

  In a double difference implementation, measurements are differenced between different satellites on one receiver in order to cancel the correlated errors. Usually one satellite is chosen as the "reference, and all others are differenced with it.

- **reference station**

  The GPS receiver which is acting as the stationary reference. It has a known position and transmits messages for the rover receiver to use to calculate its position; A ground station at a known location used to derive differential corrections. The reference station receiver tracks all satellites in view, computes their pseudoranges, corrects these for errors, and then transmits the corrections to users; A reference station is a point or site where surface stability, or tidal current constants, have been determined through accurate observations, and which is then used as a standard for the comparison of simultaneous observations at one or more subordinate stations. Certain of these are known as Continuous Operating Reference Stations (CORS), and transmit reference data on a 24-hour basis.

- **reflectance**

  Ratio of the radiant energy reflected by a body to the energy incident on it. Spectral reflectance is the reflectance measured within a specific wavelength interval.

- **reflected energy peak**

  Wavelength (0.5mm) at which maximum amount of energy is reflected from the earth's surface.

- **reflected ir**

  Electromagnetic energy of wavelengths from 0.7mm to about 3mm that consists primarily of reflected solar radiation.

- **reflection**

  The return of light or sound waves from a surface. If a reflecting surface is plane, the angle of reflection of a light ray is the same as the angle of incidence.

- **reflectivity**
  Ability of a surface to reflect incident energy.

- **refraction**
  Bending of electromagnetic rays as they pass from one medium into another when each medium has a different index of refraction; The change in direction of travel of an electromagnetic wave as it passes between two areas which have a different refractive index.

- **registration**
  Process of superposing two or more images or photographs so that equivalent geographic points coincide.

- **reject**
  To disallow a position update, usually by an operator.

- **relative**
  Applies to measurements, in a non-standard, moving reference, as opposed to fixed reference.

- **relative accuracy**
  The accuracy with which a user can measure position relative to that of another user of the same navigation system at the same time.

- **relative bearing**
  Bearing relative to heading or to the vessel; Angle from aircraft center line to bearing of the destination; Symbols: B sub R; Typical Units: rad, deg.

- **relative navigation**
  A technique similar to relative positioning except that one or both or the points may be moving. The pilot of a ship or aircraft may need to know his position relative to a harbor or runway. A data link is used to relay the error terms to the moving vessel to allow real-time navigation.

- **relative positioning**
  The determination of relative positions between two or more receivers which are simultaneously tracking the same GPS signals. One receiver is generally referred to as the Reference or Base Station, whose coordinates are known in the satellite datum. The second receiver may be stationary or moving. However its coordinates are determined relative to the Base Station. In carrier phase-based positioning this results from the determination of the baseline vector, which when added to the Base Stations coordinates generates the User's coordinates. The process of determining the relative difference in position between two locations, in the case of GPS, by placing a receiver over each site and making simultaneous measurements observing the same set of satellites at the same time. In pseudo-range-based GPS positioning, the coordinates are derived from the User receiver's observations after they have had

the differential corrections applied (either in the real-time or post-processed mode); The process of determining the relative difference in position between two marks with greater precision than that to which the position of a single point can be determined. Here, a receiver (antenna) is placed over each spot and measurements are made by observing the same satellite at the same time. This technique allows cancellation (during computations) of all errors which are common to both observers, such as satellite clock errors, propagation delays, etc.

- **relative precision**
  Precision is defined as a measure of the tendency of a set of numbers to cluster about a number determined by the set (e.g. the mean). The usual measure is the standard deviation with respect to the mean. Relative precision denotes the tendency for the various components (X, Y, Z) between one station and other stations in the network to be clustered about the adjusted values. Current custom is to express relative precision at the two-standard deviation (95% confidence) level. This may be stated in terms of a relative error ellipse or as a proportion of the separation distance (e.g. 10 ppm or 1:100,000); Precision is defined as a measure of the tendency of a set of numbers to cluster about a number determined by the set (e.g. the mean). The usual measure is the standard deviation with respect to the mean. Relative precision denotes the tendency for the various components (X, Y, Z) between one station and other stations in the network to be clustered about the adjusted values.

- **reliability**
  The ability to perform a specific function without failure under specified conditions for a given length of time.

- **relief**
  Changes in terrain; elevations or depressions in the land.

- **relief displacement**
  Geometric distortion on vertical aerial photographs. The tops of objects appear in the photograph to be radially displaced from their bases outward from the photograph's center point.

- **relief shading**
  A technique for showing the ups and downs of the land portrayed on a topographic map. The process makes land look three-dimensional by the use of graded shadow effects. Traditionally, maps are shaded as though the light source is coming from the northwest.

The use of light and dark tones to depict the form of the land; makes the map appear as if illuminated by the sun, creating highlights and shadows.

- **remapping**
  Flattening the Earth into a standard map projection. When the spherical Earth is photographed by satellites, areas lying near the outer edge of the picture are distorted. Remapping rectifies the distortion.

- **remote sensing**
  Collection and interpretation of information about an object without being in physical contact with the object; The technology of acquiring data and information about an object or phenomena by a device that is not in physical contact with it. In other words, remote sensing refers to gathering information about the Earth and its environment from a distance, a critical capability of the Earth Observing System. For example, spacecraft in low-Earth orbit pass through the outer thermosphere, enabling direct sampling of chemical species there. These samples have been used extensively to develop an understanding of thermospheric properties. Explorer-17, launched in 1963, was the first satellite to return quantitative measurements of gaseous stratification in the thermosphere. However, the mesosphere and lower layers cannot be probed directly in this way--global observations from space require remote sensing from a spacecraft at an altitude well above the mesopause. The formidable technological challenges of atmospheric remote sensing, many of which are now being overcome, have delayed detailed study of the stratosphere and mesosphere by comparison with thermospheric research advances. Some remote-sensing systems encountered in everyday life include the human eye and brain, and photographic and video cameras.

- **remote terminal (rt)**
  Term defining role of a device on a MIL-STD-1553 bus as being a slave.

- **remote/ rover receiver**
  The GPS receiver which does not know its position and needs to receive measurements from a reference station to calculate differential GPS positions. (The terms remote and rover are interchangeable.)

- **rendezvous**
  To meet with another aircraft in the air, for refueling or other mission objectives

- **rendezvous approach**
  To approach a planned rendezvous point.

- **repeater**
  Equipment which receives, amplifies, and retransmits signals in order to maintain adequate signal strength and intelligence characteristics.

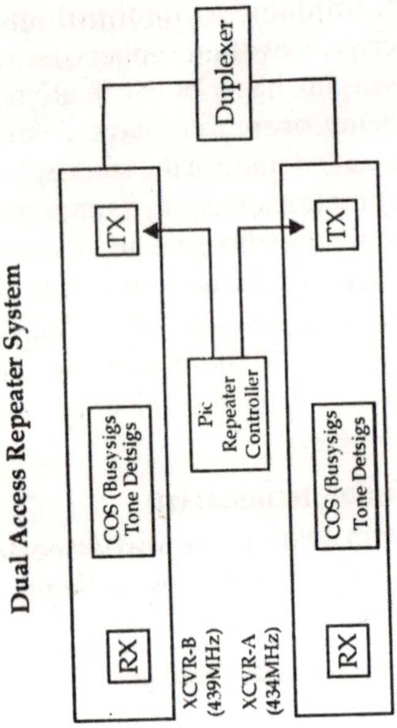

- **resampling**
  The calculation of new DN for pixels created during geometric correction of a digital scene, based on the values in the local area around the uncorrected pixels.

- **reseau marks**
  Pattern of small crosses added to photographs.

- **reservoir engineering**
  Techniques for predicting the production behaviour of oil and gas reservoirs and the optimisation of the eventual exploitation on the basis of a reservoir model, rock and fluid characteristics and flow models.

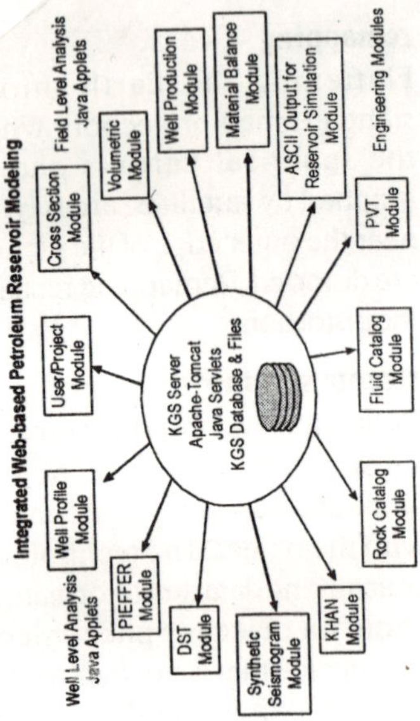

- **reset (on ins)**
  Use of external data (for example, position fix) to refine alignment of and to calibrate the inertial navigation system.

- **residual**
  In the context of measurement, the residual is the misclosure between the calculated measurements, using the position solution and actual measurements; The difference between an estimated (adjusted) value and the observed value; specifically, $v = \mu x - x$; A quality indicator for a GPS position that is

determined during the differential correction process. Indicates uncorrectable error. High residuals are not desirable.

- **resolution**

Ability to separate closely spaced objects on an image or photograph. Resolution is commonly expressed as the most closely spaced line-pairs per unit distance that can be distinguished. Also called spatial resolution.

- **resolution cell**

The smallest unit of area in an image of discrete elements. The area represented by a pixel.

- **resolution target**

Series of regularly spaced alternating light and dark bars used to evaluate the resolution of images or photographs.

- **resolving power**

A measure of the ability of individual components. and of remote sensing systems, to separate closely spaced targets.

- **reststrahlen band**

In the IR region, refers to absorption of energy as a function of silica content.

- **return**

In radar, a pulse of microwave energy reflected by the terrain and received at the radar antenna. The strength of a return is referred to as return intensity.
return-beam vidicon (rbv)A system in which images are formed on the photosensitive surface o a vacuum tube; the image is scanned with an electron beam and transmitted or recorded. Landsat 3 used a pair of RBV's to acquire images.

- **revolution**

Process of the Earth circling the sun in its orbit. Revolution determines the seasons, and the length of the year. In addition, differences in seasons occur because of Earth's inclination (tilt on its axis) of about 23.5 degrees as it revolves around the sun.

- **rf (radio frequency)**

Any frequency within the electromagnetic spectrum normally associated with radio wave propagation. Organizations such as the FCC and ITU have divided the radio frequency spectrum into subdivisions for management purposes.

- **rfid-radio frequency identification**

A method of identification using radio frequency. For objects, animals, or people and does not require direct contact or line-of-sight scanning.

- **rhumb line**

A line that passes through all meridians at the same angle. When drawn on a Mercator chart, the rhumb line is a straight line, because the Mercator chart is a distortion of a spherical globe on a flat surface. The rhumb line results in a longer course than a great circle route; A line of the surface of the earth that crosses all meridians at a constant angle.

- **right ascension of ascending node (aka =, raan or ra of node)**
One of six Keplerian elements, it indicates the rotation of the orbit plane from some reference point. Two numbers orient an orbital plane in space; inclination is the first, this is the second. After specifying inclination, an infinite number of orbital planes are possible. The intersection of the equatorial plane and the orbital plane must be specified by a location on the equator that fully defines the orbital plane. The line of nodes occurs in two places. However, by convention, only the ascending node (where the satellite crosses the equator going from south to north) is specified. The descending node (where the satellite crosses the equator going from north to south) is not. Because the Earth spins, conventional latitude and longitude points are not used to separate where the lines of node occur. Instead, an astronomical coordinate system is used, known as the right-ascension/declination coordinate system, which does not spin with the Earth. Right ascension of ascending node is an angle, measured at the center of the Earth, from the vernal equinox to the ascending node. For example, draw a line from the center of the Earth to the point where the satellite crossed the equator (going from south to north). If this line points directly at the vernal equinox, then RAAN = 0 degree.

- **rinex**
Receiver independent exchange format. A set of standard definitions and formats to promote the free exchange of GPS data and facilities the use of data from any GPS receiver with any software package. The format includes definitions for three fundamental GOS observables: time, phase, and range.

- **rinex-receiver-independent exchange**
A GPS data format, which includes provisions for pseudorange, carrier-phase, and Doppler observations.

- **ringing**
Fringe-like artefacts produced at edges by some forms of spatial-frequency filtering.

- **ring-laser gyro (rlg)**
A gyroscope based on a laser beam instead of a rotating mass, providing to same data as a gyroscope.

- **rkp-relative kinematic positioning**
  Improves GPS accuracy by using a combination of differential GPS (DGPS) correction data, transmitting GPS signal phase information and ambiguity resolution techniques via statistical tests-possibly with processing in real-time (real-time kinematic positioning, RTK).

- **rms**
  Root Mean Square deviation is also known as the population standard deviation is given by the formula; Equation Where: μ is the estimated (adjusted) value for x and N is the number of observations; Literally "root mean square." A DC voltage that will produce the same heating effect (power output in Watts) as the AC voltage. For a sine wave, the RMS value is equal to 0.707 times the peak value of an AC voltage. Example: divide Peak-to-Peak by 2 (or in half) and multiple by 0.707 = RMS voltage.

- **rnp**
  Required Navigation Performance standards.

- **road pricing**
  Use of smart card technology or simple tags to charge road users based on criteria such as demand, congestion, time, or distance.

- **roaming**
  The use of a wireless phone outside of the "home" service area defined by a service provider. Higher per-minute rates are usually charged for calls made or received while roaming. Long distance rates and a daily access fee may also apply.

- **rods**
  The receptors in the retina that are sensitive to brightness variations.

- **roll**
  Rotation of an aircraft that causes a wing-up or wing-down attitude.

- **roll angle (roll attitude)**
  The angle between the horizontal and the lateral axis of the craft. See: bank, list.

- **roll compensation system**
  Component of an airborne scanner system that measures and records the roll of the aircraft. This information is used to correct the imagery for distortion due to roll.

- **roll cue**
  Flight director cue to control roll; in fixed-wing aircraft, a wheel cue; in rotary-wing aircraft, a lateral cyclic cue.

- **roll rate**
  Rate of change of roll; time derivative of roll; Symbols: q; Symbols: phi dot; Typical Units: rad/s, deg/s; Dimensions: 1/Time;

- **rom**
  Read Only Memory refers to the computer memory chips that

contain information the computer uses (along with system files) throughout the system, including the information it needs to get itself started. Information in ROM is permanent; it doesn't vanish when the power is turned off.

- **root mean square (rms)**
  A statistical measure of data; the root of the mean of the square; for variables with mean of zero, the standard deviation is equal to the rms.

- **root sum square (rss)**
  A statistical measure of data; the root of the sum of the square; for a vector, its length is equal to the rss of its scalar elements.

- **rough criterion**
  In radar, the relationship between surface roughness, depression angle, and wavelength that determines whether a surface will scatter the incident radar pulse in a rough or intermediate fashion.

- **roughness**
  In radar, the average vertical relief of a small-scale irregularities of the terrain surface. Also called surface roughness.

- **route**
  A planned course of travel, usually composed of more than one navigation leg; A group of waypoints entered into the GPS receiver in the sequence you desire to navigate them; Consists of two or more waypoints combined in a course of travel. It provides the automatic capability to navigate through several waypoints, without having to reprogram the unit after arriving at each one. Once programmed into the GPS unit, the route provides the option of navigating forward through the waypoints or in reverse order.

- **router**

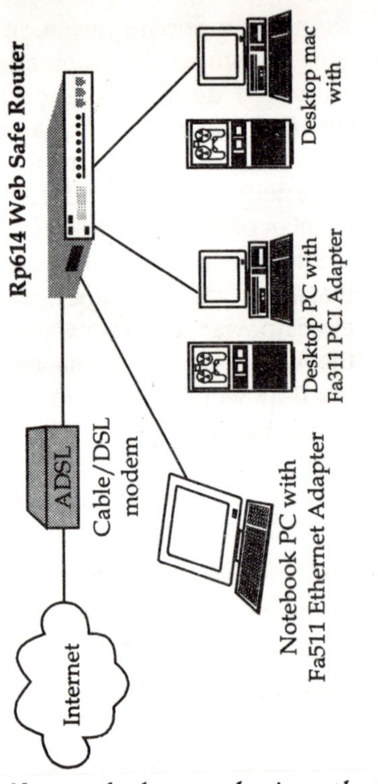

Network layer device that determines the optimal path along which network traffic should be forwarded. Routers forward packets from one network to another based on network layer information.

- **rov**

  Remotely Operated Vehicle. Unmanned submersible launched from a ship and equipped with measuring and manipulation equipment. A cable to the mother-ship provides power, video and data communication.

- **rover**

  Any mobile GPS receiver collecting data during a field session. The receiver's position may be computed relative to another, stationary GPS receiver at a Base Station. May also be referred to as the Mobile Receiver.

- **rs232**

  The electronic voltage and timing interface definition for the connection of electronic equipment.

- **rt-20**

  NovAtel's Double Differencing Technology for real-time kinematic (RTK) carrier phase floating ambiguity resolution.

- **rtcm**

  Radio Technical Commission for Maritime Services Commission set up to define a differential data link to relay GPS correction messages from a monitor station to a field user. RTCM SC-104 recommendations is the defacto standard for differential GPS correction transmission. It defines the correction message format and 16 different correction message types.

- **rtcm sc-104**

  The most common format used in the transmission of DGPS differential corrections, including those sent by Coast Guard beacons; The special committee of the Radio Technical Commission for Maritime Services that developed recommended standards for DGPS.

- **rtcm-radio technical commission for maritime services**

  The commission set up to define a differential data link to relay GPS correction messages from a monitor station to a field user. RTCM SC-104 recommendations define the correction message format and 16 different correction message types.

- **rtk otf**

  Real-Time Kinematic (RTK) On-The-Fly (OTF) positioning. A DGPS technique for high precision positioning of moving objects

- **rtk-real time kinematic**

  Land survey based on a differential use of carrier phase measurements of the GPS.

Glonass and/or Galileo signals where a single reference station provides the real-time corrections of even to a centimetre level of accuracy.

- **rudder**

  A control surface on fixed-wing aircraft, usually mounted at aft end of the fuselage sticking up (like a dorsal fin), that controls yaw (heading), and is controlled by the pedals; Symbols: delta sub R; Typical Units: rad, deg;

- **sa**

  Selective Availability. A policy adopted by the Department of Defense to introduce some intentional clock noise into the GPS satellite signals thereby degrading their accuracy for civilian users. This policy was discontinued as of May 1, 2000 and now SA is turned off

- **saddle**

  A dip along the top of a ridge.

- **samii**

  Stratospheric Aerosol Measurement experiment, carried by Nimbus-7.

- **sampling**

  The process of obtaining a sequence of discrete digital values from a continuous sequence of analog data.

- **sampling frequency**

  Rate of a periodic process; 1/T where T is the period; Symbols: f sub s; Typical Units: Hz; Dimensions: 1/Time.

- **sams**

  Stratospheric and Mesospheric Sounder, carried by Nimbus-7.

- **sar-synthetic aperture radar**

  A method of active illumination remote sensing in which the instrument emits microwaves, and uses the range and Doppler shift of the returns to locate ground points. SAR instruments are mounted on the ERS 1 and 2 satellites.

- **sar (specific absorption rate)**

  Measures the maximum quantity of radiation absorbed by one Kilogram of tissue from a cell phone. Nokia, Motorola and Ericsson are working to establish international benchmark standards.

- **sarsat**

  Search and Rescue Tracking System carried on NOAA polar-orbiting satellites that receives emergency signals from persons in distress. The satellites transmit these signals to ground receiving stations in the U.S. and overseas. Signals are forwarded to the nearest rescue coordination center which computes the location from which the emergency signals came and provides the coordinates of the emergency site to a rescue team.

- **satellite**

A radio relay station that orbits the earth. A complete satellite communications system also includes earth stations that communicate with each other via the satellite. The satellite receives a signal transmitted by an originating earth station and retransmits that signal to the destination earth station(s). Satellites are used to transmit telephone, television and data signals originated by common carriers, broadcasters and distributors of cable TV program material.

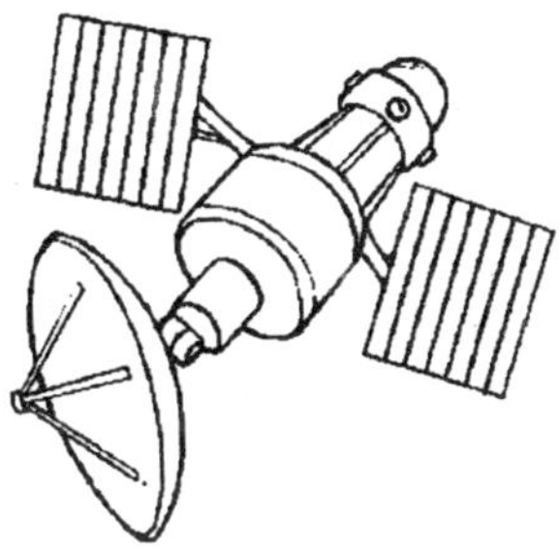

- **satellite constellation**

The arrangement in space of a set of satellites; The group of GPS satellites from which data is used to determine a position.

- **satellite home viewer improvement act of 1999 (shvia)**

An Act modifying the Satellite Home Viewer Act of 1988. SHVIA permits satellite companies to provide local broadcast TV signals to all subscribers who reside in the local TV station's market. SHVIA also permits satellite companies to provide "distant" network broadcast stations to eligible satellite subscribers.

- **satellite image mapping (sim)**

SIM is a product of remote sensing where discrete blocks of orbital photography are combined into a comprehensive whole and geocoded, or computer-linked to specific Mercator, Lambert Conformal, or other types of projections that include a scale factor and reference geoid, with each pixel related to a specific latitude and longitude.

- **satellite master antenna television (smatv)**

A satellite dish system used to deliver signals to multiple dwelling units (e.g., apartment buildings and trailer parks).

- **satellite pass**

Segment of orbit during which the satellite "passes" nearby and in range of a particular ground station.

- **satellite positioning**

A procedure by which satellites are used to locate precise objects or particular points on Earth.

- **satellite revolution**

The time from one perigee (the point of an elliptical orbit path where a satellite is closest to Earth) to the next.

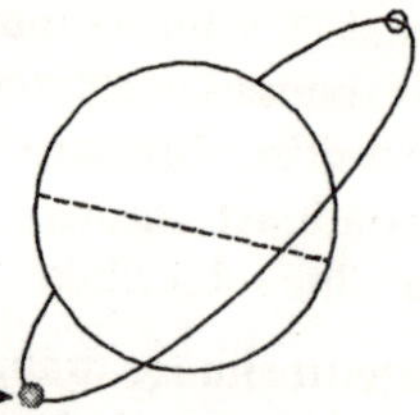

- **satellite status display**

An information screen that shows technical data about each satellite in view. Information includes receiver channel numbers; actual satellite I.D. numbers; status of satellite tracking (T) or searching (S); satellite elevations and azimuths; signal-to-noise ratios (higher the number, the better); and dilution of precision ratings (GDOP is most important; smaller the number, the better).

- **satellite-based augmentation system (sbas)**

A type of geo-stationary satellite system that improves the accuracy, integrity, and availability of the basic GPS signals. This includes WAAS, EGNOS, and MSAS; A Satellite-Based Augmentation System is any system that employs a network of geostationary satellites and ground stations to enhance the performance of a Global Navigation Satellite System (GNSS) by providing signal corrections and additional satellites with known and constant positions. Three SBAS are currently in operation around the world, covering specific geographic regions; North America's Wide Area Augmentation System (WAAS), Europe's European Global Navigation Overlay System (EGNOS) and Asia's Multi-transportation Satellite Augmentation System (MSAS); Networks of ground relay stations and geostatic satellites designed to receive satellite navigation signals and transmit corrected time and distance measurements that greatly improve accuracy. Observation and relay stations have been set at known positions all over the world, while their geostatic satellites continuously maintain a fixed position above the Earth. Using these known values for distance, SBAS corrects satellite navigation signals for atmospheric delays, incorrect satellite positioning and poor geometry, sometimes caused by inline or close alignment of satellites, increasing accuracy in specific regions. SBAS is vital to providing the reliability and precision required by aviation and other precision-critical applications.

- **satnav**

A local term referring to use of the older TRANSIT system for satellite navigation. One major difference between TRANSIT and GPS is that the TRANSIT satellites are in low-altitude polar orbits with a 90-minute period.

• **saturation**
In the IHS system, represents the purity of color. Saturation is also the condition where energy flux exceeds the sensitivity range of a detector; A condition that exists when a further change in the input produces no additional output (a saturated amplifier).

• **savable plot trails**
The capability to save the actual trail created on the plotter display, thereby enabling the GPS user to either backtrack the course immediately...or retrace the trip at a later time.

• **scalar feed**
A type of horn antenna feed which uses a series of concentric rings to capture signals that have been reflected toward the focal point of a parabolic antenna.

• **scale**
Ratio of distance on an image to the equivalent distance on the ground; The distance between two points on a map as they relate to the distance between those same points on the earth

• **scan line**
Narrow strip on the ground that is swept by IFOV of a detector in a scanning system.

• **scan skew**
Distortion of scanner images caused by forward motion of the aircraft or satellite during the time required to complete a scan.

• **scanner**
An imaging system in which the IFOV of one or more detectors is swept across the terrain; A system that optically scans its detector(s) across a scene and records or stores the data in a two-dimensional format to form an image; A radio receiver that moves across a wide range of radio frequencies and allows audiences to listen to any of the frequencies.

• **scanner distortion**
Geometric distortion that is characteristic of cross-track scanner images.

• **scanning radiometer**
An imaging system consisting of lenses, moving mirrors, and solid-state image sensors used to obtain observations of the Earth and its atmosphere. Scanning radiometers, which are the sole imaging systems on all current operational weather satellites, have far better long-term performance than the vidicon TV camera tubes used with earlier spacecraft.

• **scattering**
Multiple reflections of electromagnetic waves by particles or surfaces; The process

by which electromagnetic radiation interacts with and is redirected by the molecules of the atmosphere, ocean, or land surface. The term is frequently applied to the interaction of the atmosphere on sunlight, which causes the sky to appear blue (since light near the blue end of the spectrum is scattered much more than light near the red end).

- **scattering coefficient curves**
  Display of scatterometer data in which relative backscatter is shown as a function of incidence angle.

- **scatterometer**
  Nonimaging radar device that quantitatively records backscatter of terrain as a function of incidence angle.

- **scene**
  Area on the ground that is covered by an image or photograph.

- **scotopic vision**
  Vision under conditions of low illumination, when only the rods are sensitive to light. Visual acuity under these conditions is highest in the blue part of the spectrum.

- **scrambler**
  A device used to electronically alter a signal so that it can only be viewed or heard on a receiver equipped with a special decoder.

- **screaming eagles**
  Cloud pattern so named because some observers maintain they can see the head of an eagle facing west in these cloud patterns. The pattern is similar to a comma, only the pattern is disorganized and not solid. Weather associated with screaming eagles consists of rain showers and gusty surface winds up to about 25 knots. The eagles can intensify and enlarge when moving into areas east of troughs; in that case, intense thunderstorms can develop. Screaming eagles are common in the Pacific Ocean between Hawaii and the equator, and are uncommon in the western Atlantic.

- **sdh (synchronous digital hierarchy)**
  An international digital telecommunications network hierarchy which standardizes transmission around the bit rate of 51.84 megabits per second, which is also called STS-1. Multiples of this bit rate comprise higher bit rate streams. SONET is the American version of SDH.

- **sea breeze**
  Local coastal wind that blows from the ocean to land. Sea breezes usually occur during the day, because the heating differences of land and sea cause pressure differences. Cooler, heavier air from the sea moves in to replace rising warm air on the coastline.

• **sea level**
The datum against which land elevation and sea depth are measured. Mean sea level is the average of high and low tides.

• **search and rescue**
International satellite-aided search and rescue project. COSPAS/SARSAT satellites monitor the entire surface of the Earth, and transmit distress signals to special ground receiving stations. The receiving stations compute the location of the signal, and notify the nearest rescue coordination center. Satellite search has cut recovery time from days to hours, and has aided downed air planes, capsized boats, and persons in other emergencies.

• **search pattern**
Basic guidance mode; provides lateral steering guidance to fly an expanding square search pattern, creeping line search pattern, or sector search pattern.

• **search the sky**
A message shown when a GPS receiver is gathering satellite almanac data. This data tells the GPS receiver where to look for each GPS satellite.

• **secam**
A color television. system developed by the French and used in the USSR. Secam operates with 625 lines per picture frame and 50 cycles per second, but is incompatible in operation with the European PAL system or the U.S. NTSC system; A French developed color television standard which is common in Francophile countries and the former Soviet Union. The system operates with a screen density of 625 scanlines per picture and 50 cycles per second but inverts the signal making it incompatible with the PAL and NTSC video standards.

• **second**
A unit of measurement; one minute contains 60 seconds; one degree contains 3600 seconds.

• **second-order filter**
A smoothing filter in which the output follows the input, only more slowly; It is usually implemented in software as a difference equation of period T. When the second-order filter is used in avionics, it is commonly to smooth data, and to wash out transients at mode change. Usually, a first-order filter suffices, and it being less expensive, is chosen over a second-order filter. Typical values for omega sub n, a, and b are 0.1-2 rad/s, and zeta 0.1-0.9. It should also be noted that two first-order filters can be chained together to form a second-order filter that is critically damped or over damped. When implementing a second-order filter on normalized variables, such as angles, the

discontinuities require special treatment.

- **sector search**
A pattern of concentric arcs followed for searching the ground from an aircraft.

- **secular motion**
That portion of crustal motion which is continuous and at a constant velocity. Secular motion is uniformly predictable over time and is independent of any seismic events.

- **secular variation**
Complex fluid motion in the outer core of the Earth that causes the magnetic field to change slowly with time.

- **secure digital**
A type of technology for mass storage devices in the form of a type of memory media with a compact design and is compatable with different·digital camera brands and MP3 devices.

- **see-thru® technology**
A Garmin exclusive technology which allows the various Garmin fishfinders to hear both weak and strong signals simultaneously so as to identify fish returns under the toughest conditions: suspended in thermoclines or even hiding near structures.

- **selected**
Measured data picked from one of many sensors; Selecting is the process of choosing the "best" parameter from multiple copies of that parameter, from multiple, identical devices or similar devices.

- **selected waypoint**
he waypoint currently selected to be the point toward which the vessel is travelling. Also called to waypoint, destination or destination waypoint.

- **selective availability**
SA A DoD program to control the accuracy of pseudorange measurements, whereby the user receives a false pseudorange which is in error by a controlled amount. Differential and phase differencing techniques can reduce these effects for local applications.

- **self-contained navigation aid**
An aid which consists only of facilities carried by the vehicle.

- **semi-circle**
A measure of angle, 1 semi-circle = pi rad = 180 deg; angles from physical devices are often reported in semi-circles in order to compress data

- **semi-major axis**
One half of the major axis of an ellipse.

- **semi-major axis (aka a)**
One of the six Keplerian elements, it indicates the size of an orbit. The semi-major axis is one-half of the

longest diameter of an orbital ellipse, e.g., one-half of the distance between the apogee and perigee of an Earth orbit. (The semi-major axis is related to the orbital period and mean motion by Kepler's third law.

- **sense**
  The pointing direction of a vector representing some navigation parameter.

- **sensitivity**
  This is a manual adjustment that allows users to fine-tune the fish finder to see the most underwater detail with the least visual noise. For anglers who'd rather spend more time fishing than adjusting, Advanced Signal Processing (ASP™) does this automatically, and is a feature of all Lowrance sonar units.

- **sensor**
  A device that measures, receives, or generates data, for example, an INS, a FLIR, a map; Device that receives electromagnetic radiation and converts it into a signal that can be recorded and displayed as either numerical data or an image; Sensors aboard satellites obtain information about features and objects on Earth by detecting radiation reflected or emitted in different bands of the electromagnetic spectrum. Analyzing the transmitted data provides valuable scientific information about Earth. Weather satellites commonly carry radiometers, which measure radiation from snow, ice, clouds, and bodies of water. Spaceborne radars are used for Earth observations, bouncing radar waves off land and ocean surfaces to study sea-surface conditions, ice thickness, and land surface features. A wind scatterometer is a special type of radar designed to measure ocean surface winds indirectly by bouncing signals off the water and measuring them from various angles. Infrared (IR) detectors measure heat generated by Earth features in the IR band of the spectrum. Photographic reconnaissance sensors in their simplest form are large telescope-camera systems used to view objects on Earth's surface. The bigger the lens, the smaller the object that can be detected. Camera-telescope systems now incorporate all sorts of sophisticated electronics to produce better images, but even these systems need cloudless skies, excellent lighting, and good color contrast between objects and their surroundings to detect objects the size of a basketball. Some of the satellites produce film images that must be returned to Earth, but a more convenient method is to record the image as a series of digital code numbers, then reconstruct the image from

the electronic code using a computer at a ground station.

- **sensor calibration**
  The relationship between input and output for a given measurement.

- **sep**
  Spherical Error Probable, a statistical measure of precision defined as the 50th percentile value of the three-dimensional position error statistics. Thus, half of the results are within a 3-D SEP value.

- **sep-spherical error probable**
  The radius of a sphere within which there is a 50 per cent probability of locating a point or being located.

- **sequential receiver**
  A GPS receiver in which the number of satellite signals to be tracked exceeds the number of available hardware channels. Sequential receivers periodically reassign hardware channels to particular satellite signals in a predetermined sequence.GPS receivers vary widely in accuracy because of the expense of adding more radio receivers needed to tune in more satellites. For instance, early consumer-grade receivers typically included six to eight receivers for the L1 C/A signal.

- **serial communication**
  The sequential transmission of the signal elements of a group representing a character or other entity of data. The characters are transmitted in a sequence over a single line, rather than simultaneously over two or more lines, as in parallel transmission. The sequential elements may be transmitted with or without interruption.

- **service plan**
  The rate plan you select when choosing a wireless phone service. A service plan typically consists of a monthly base rate for access to the system and a fixed amount of minutes per month.

- **service provider**
  A telecommunications provider that owns circuit switching equipment.

- **set-surface elevation table**
  A portable mechanical leveling device for measuring the relative elevation of wetland sediments.

- **sfd - stauration flux density**
  The power required to achieve saturation of a single repeater channel on the satellite.

- **shared hub**
  A satellite communications operations center which is shared among a number of separate network users; often used for VSAT operations.

- **sidelap**
  Extent of lateral overlap between images acquired on adjacent flight lines.

- **sidelobe**
  Off-axis response of an antenna.
- **side-looking airborne radar (slar)**
  An airborne side scanning system for acquiring radar images.
- **sidereal day**
  Time between two upper transits of the vernal equinox. A sidereal day is approximately four minutes shorter than a solar day.
- **sidereal time**
  Sidereal time is defined by the hour angle of the vernal equinox. Taking the mean equinox as the reference yields true or apparent Sidereal Time. Neither Solar nor Sidereal Time are constant, since angular velocity varies due to fluctuations caused by the Earth's polar moment of inertia as exerted through tidal deformation and other mass movements.
- **side-scanning sonar**
  Active system for acquiring images of the seafloor using pulsed sound waves.
- **side-scanning system**
  A system that acquires images of a strip of terrain parallel with the flight or orbit path but offset to one side.
- **signal**
  Information recorded by a remote sensing system.
- **signal data converter (sdc)**
  A device that converts unique signals to a standard protocol, usually MIL-STD-1553B.
- **signal to noise radio (s/n)**
  The ratio of the level of the signal carrying real information to that carrying spurious information as a result of defects in the system; The ratio of the signal power and noise power. A video S/N of 54 to 56 dB is considered to be an excellent S/N, that is, of broadcast quality. A video S/N of 48 to 52 dB is considered to be a good S/N at the headend for Cable TV.
- **signal to noise ratio (snr)**
  The ratio of incoming signal strength to the amount of interfering noise as measured in decibels on a logarithmic scale
- **silver halide**
  Silver salts that are especially sensitive to visible light and convert to metallic silver when developed.
- **simplex operation**
  That mode of operation in which communication between two points occurs in only one direction at a time. Contrast with half duplex or duplex operation.
- **simplex transmission**
  Capability for transmission in only one direction between sending station and receiving station.

- **simplified gps receiver block diagram**

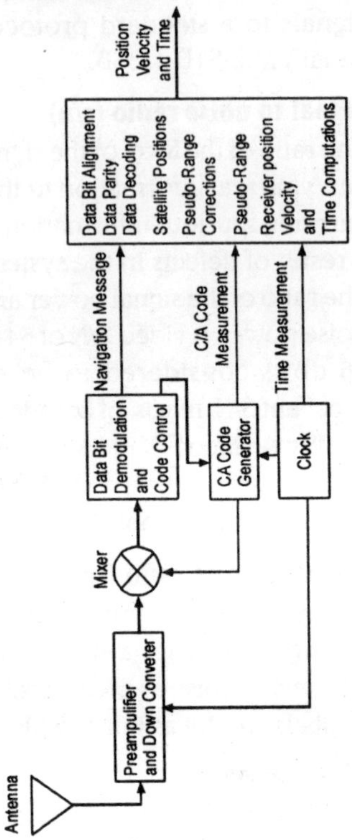

- **simultaneous measurements**

  Measurements referenced to time-frame epochs which are either exactly equal, or else so closely spaced in time that the time misalignment can be accommodated by correction terms in the observation equation, rather than by parameter estimation.

- **sinex**

  Solution Independent Exchange format. A solution output format recently developed by geodesists to permit the exchange of solution information between organisations, from which the original normal equation systems for precise GPS adjustments can be reconstructed. These reconstructed equation systems can be combined with other normal equation systems to create new GPS baseline solutions.

- **single difference**

  A GPS observable formed by arithmetically differencing carrier phases that are simultaneously measured by a pair of receivers tracking the same satellite, or by a single receiver tracking a pair of satellites. The between-satellite single-difference procedure essentially removes all receiver clock errors. Although primarily used with carrier phases, the procedure can also be applied to pseudoranges.

- **single-channel-per-carrier (scpc)**

  A method used to transmit a large number of signals over a single satellite transponder.

- **sink**

  The process of providing storage for a substance. For example, plants--through photosynthesis--transform carbon dioxide in the air into organic matter, which either stays in the plants or is stored in the soils. The plants are a sink for carbon dioxide.

- **sir**
  Shuttle Imaging Radar, synthetic-aperture radar experiments carried aboard the NASA Space Shuttle in 1981 and 1984.

- **situational awareness**
  Situational awareness provides the pilots with information relative to the current surroundings, such as other aircraft or threats in the immediate area.

- **skew**
  An adjustment that compensates for slight variance in angle between identical senses of polarity generated by two or more satellites.

- **skimmer®**
  Available in a variety of operating frequencies, Skimmer transducers feature a low profile for high-speed performance...even up to 70 mph! The patented design helps prevent cavitation and other interference that affects sonar performance. Finely tuned to provide exceptional echo sounding, transom-mounted models feature a kick-up bracket for protection against impact with debris and obstructions.

- **skyfix**
  DGPS positioning system originating from Thales GeoSolutions.

- **skylab**
  The first U.S. space station, launched unmanned in May 1973 and soon after occupied in succession by three crews through November 1973.

- **skylight**
  Component of light that is strongly scattered by the atmosphere and consists predominantly of shorter wavelengths.

- **slamming**
  The term used to describe what occurs when a customer's long distance service is switched from one long distance company to another without the customer's permission. Such unauthorized switching violates FCC rules.

- **slant range**
  In radar, an imaginary line running between the antenna and the target; The length of the path between a communications satellite and an associated earth station.

- **slant-range distance**
  Distance measured along the slant range.

- **slant-range distortion**
  Geometric distortion of a slant-range image.

- **slant-range image**
  In radar, an image in which objects are located at positions corresponding to their slant-range distances from the aircraft path. On slant-range images, the scale in the range direction is compressed in the near-range region.

- **slats**
A control surface on fixed-wing aircraft, usually mounted to the aft edge of the wings, that extends the wing to provide added lift at low speeds.

- **sleeve**
An attachment for a PDA which allows you to use items like network or modem cards, PCMCIA or CF cards, digital cameras, a GPS/GPRS phone, or an extra battery pack with a PDA.

- **slope distance**
The three-dimensional vector distance from station one to station two. The shortest distance (a chord) between two points.

- **slot**
That longitudinal position in the geosynchronous orbit into which a communications satellite is "parked". Above the United States, communications satellites are typically positioned in slots which are based at two to three degree intervals.

- **slow switching channel**
A sequencing GPS receiver channel that switches too slowly to allow the continuous recovery of the data message.

- **smg**
Speed Made Good. Marine term Giving speed to waypoint taking into effect the course you are steering.

- **smirr**
Shuttle Multispectral Infrared Radiometer, a non-imaging spectroradiometer carried by the NASA Space Shuttle covering ten narrow wavebands in the 0.5-2.4 m range.

- **smmr**
Scanning Multichannel Microwave Radiometer, carried by Nimbus-7.

- **smooth criterion**
In radar, the relationship between surface roughness, depression angle, and wavelength that determines whether a surface will scatter the incident radar pulse in a smooth or intermediate fashion.

- **smoothed**
Data that is the result of conditioning a signal with a simple filter.

- **smoothing filter**
A filter to reduce quick changes of a signal by attenuating high frequencies.

- **sms**
Short Message Service (also called text messaging): A function available on digital GSM networks allowing text messages to be sent and received via the network operator to a mobile phone, or from the Internet, using an "SMS gateway" website.

- **snafu**
Situation Normal, All Fouled UP. Military term.

- **sng**
  Satellite news gathering usually with a transportable uplink truck.

- **snow**
  A form of noise picked up by a television receiver caused by a weak signal. Snow is characterized by alternate dark and light dots appearing randomly on the picture tube. To eliminate snow, a more sensitive receive antenna must be used, or better amplification must be provided in the receiver (or both).

- **software**
  The programs, data, or routines used by a computer, distinguished from the physical components (e.g., hardware); Software usually refers to a set of advanced computer modules that allow the user to plan efficient surveys, organize and acquire satellite navigation data, verify and download data, process and analyze measurements, perform network adjustments, and report and archive the final results.

- **sog**
  Speed Over Ground. Marine term to tell you to ignore tidal and current effects.

- **soit**
  FAA's Satellite Operational Implementation Team.

- **solar backscatter ultraviolet radiometer (sbuv)**
  Instrument that measures the vertical distribution and total ozone in the Earth's atmosphere. Data is used for the continuous monitoring of ozone distribution to estimate long-term trends. SBUV instruments are flown on NOAA polar-orbiting satellites.

- **solar constant**
  Aka total solar irradiance. The constant expressing the amount of solar radiation reaching the Earth from the sun, approximately 1370 watts per square meter. It is not, in fact, truly constant and variations are detectable.

- **solar cycle**
  Eleven-year cycle of sunspots and solar flares that affects other solar indexes such as the solar output of ultraviolet radiation and the solar wind. The Earth's magnetic field, temperature, and ozone levels are affected by this cycle.

- **solar day**
  The Solar Day contains exactly 24 hours (1440 minutes). During the Solar Day, the Earth rotates slightly more than 360 degrees.

- **solar outage**
  Solar outages occur when an antenna is looking at a satellite, and the sun passes behind or near the satellite and within the field of view of the antenna. This field of view is usually wider than the beam width. Solar outages can be exactly predicted as to the timing for each site.

- **solar radiation**

Energy received from the sun is solar radiation. The energy comes in many forms, such as visible light (that which we can see with our eyes). Other forms of radiation include radio waves, heat (infrared), ultraviolet waves, and x-rays. These forms are categorized within the electromagnetic spectrum.

- **solar wind**

A continuous plasma stream expanding into interplanetary space from the sun's corona. The solar wind is present continuously in interplanetary space. After escaping from the gravitational field of the sun, this gas flows outward at a typical speed of 400 km per second to distances known to be beyond the orbit of Pluto. Besides affecting Earth's weather, solar activity gives rise to a dramatic visual phenomena in our atmosphere. The streams of charged particles from the Sun interact the Earth's magnetic field like a generator to create current systems with electric potentials of as much as 100,000 volts. Charged electrons are energized by this process, sent along the magnetic field lines towards Earth's upper atmosphere, excite the gases present in the upper atmosphere and cause them to emit light which we call the auroras. The auroras are the northern (aurora borealis) and southern (aurora Australis) lights.

- **solid state power amplifier (sspa)**

A high power amplifier using semiconductors to amplify the signal.

- **sonar**

Acronym for sound navigation ranging. Sonar is an active form of remote sensing that employs sonic energy to image the seafloor; A system using transmitted and reflected underwater sound waves to detect and locate submerged objects or measure the distance to the floor of a body of water. This technology is used in Garmin fishfinders and sounder products.

- **sonet (synchronous optical network)**

A synchronous digital network that employs optical fibers (exclusively). Each facility conforms to standards that include electrical and optical speeds, and frame formats. SONET carries circuit-switched data in frames at speeds in multiples of 51.84 megabits per second. SDH is the international term for SONET.

- **sounder**

A special kind of radiometer that measures changes in atmospheric temperature with height, as well as the content of various chemical species in the atmosphere at various levels. The High Resolution Infrared Radiation Sounder (HIRS), found on NOAA polar-orbiting satellites, is a passive instrument.

- **space environment monitor (sem)**
  Instrument that measures the condition of the Earth's magnetic field and the solar activity and radiation around the spacecraft, and transmits these data to a central processing facility. NOAA polar-orbiting and geostationary satellites both carry SEMs.

- **space physics**
  Scientific study of magnetic and electric phenomena that occur in outer space, in the upper atmosphere of the planets, and on the sun.

- **space segment**
  The space-based component of the GPS system (i.e. the satellites). The satellites may be differentiated into various groups, e.g. the Block II, Block IIA, Block IIR, and Block IIF satellites. The part of the whole GPS system that is in space, i.e. the satellites. A term which describes the portion of the total communications satellite system which is physically located in orbit around the earth.

- **space shuttle**
  NASA's manned, recoverable spacecraft designed to be used as a launch vehicle for Earth-orbiting experiments and as a short-term research platform.

- **space station**
  A planned series of three polar-orbiting, sun-synchronous satellites to be launched by NASA, the European Space Agency, and the Japanese Space Agency in the 1990s. They will carry a large range of remote-sensing devices.

- **space vehicle id (sv)**
  Sometimes used as SVID. A unique number assigned to each satellite for identification purposes. The 'space vehicle' is a GPS satellite.

- **spacelab**
  A manned laboratory module built by the European Space Agency (ESA) that accommodates dozens of experiments on each flight, mainly in the categories of materials science and life science.

- **spacelink**
  NASA electronic database for educators, with information stored on a computer at the Marshall Space Flight Center. Via computer, educators communicate with NASA education specialists and access the following menus: current NASA news, aeronautics research, U.S. Space Program historical information, aerospace research in the 1980s and beyond, overviews of NASA and its Centers, NASA educational services, classroom materials, and

space program spin-offs. The computer access number is 205-895-0028, the data word format is 8 data bits, no parity, and 1 stop bit--300, 1200, or 2400 baud modem required. Callers with Internet access may reach NASA Spacelink at: spacelink.msfc.nasa.gov.

- **spatial decorrelation**
  The distance between the user and the reference station. When calculating differential corrections, the greater the distance between the two, the greater the error of corrections.

- **spatial-frequency filtering**
  The analysis of the spatial variations in DN of an image and the separation or suppression of selected frequency ranges.

- **specific heat**
  The ratio of the heat capacity of unit mass of a material to the heat capacity of unit mass of water.

- **spectral band**
  A finite segment of wavelengths in the electromagnetic spectrum.

- **spectral hue**
  A hue which is present in the spectral range of white light analysed by a prism or diffraction grating.

- **spectral reflectance**
  Reflectance of electromagnetic energy at specified wavelength intervals.

- **spectral sensitivity**
  Response, or sensitivity, of a film or detector to radiation in different spectral regions.

- **spectral signature**
  The variation of reflectance with wavelength for an object.

- **spectral vegetation index**
  An index of relative amount and vigor of vegetation. The index is calculated from two spectral bands of AVHRR imagery.

- **spectrometer**
  Device for measuring intensity of radiation absorbed or reflected by a materiel as a function of wavelength.

- **spectroradiometer**
  A device which measures the energy reflected or radiated by materials in narrow EM wavebands.

- **spectrum**
  Continuous sequence of electromagnetic energy arranged according to wavelength or frequency.
  1. The series of colored bands diffracted and arranged in the order of their respective wave lengths by the passage of white light through a prism or other diffracting medium and shading continuously from red (produced by the longest visible wave) to violet (produced by the shortest visible wave).

2. Any of various arrangements of colored bands or lines, together with invisible components at both ends of the spectrum, similarly formed by light from incandescent gases or other sources of radiant energy, which can be studied by a spectrograph.
3. In radio, the range of wave lengths of radio waves, from 3 centimeters to 30,000 meters, or of frequencies of radio waves, from 10 to 10,000,000 kilocycles. Also radio spectrum.
4. The entire range of radiant energies.

- **specular**
Refers to a surface that is smooth with respect to the wavelength of incident energy.

- **speed cue**
Flight director cue to control speed; in fixed-wing aircraft, a throttle cue; in rotary-wing aircraft; a longitudinal cyclic cue

- **speed error**
A basic output from guidance to flight director, indicating the difference between actual speed and desired speed; Symbols: DELTA V; Typical Units: ft/s,kt; Dimensions: Length / Time.

- **speed made good (smg)**
Marine term Giving speed to waypoint taking into effect the course you are steering.

- **speed map**
A computer generated map compiled from data received from vehicle detection data which displays various colors for differing road conditions. This data gives a real-time picture of the current traffic speeds.

- **speed over ground**
The speed at which the GPS unit and the person operating it are moving with respect to the earth's surface (ignoring tidal and current activity); The actual speed the GPS unit is moving over the ground. This may differ from airspeed or nautical speed due to such things as head winds or sea conditions. For example, a plane that is going 120 knots into a 10-knot head wind will have a SOG of 110 knots.

- **speed pulse generator kit**
Some older vehicles (mainly pre-1980s) do not have vehicle speed sensor wires. You'll need a special speed pulse generator kit to adapt these vehicles to accept a navigation system.

- **speed sensor lead**
This lead connects to your vehicle's speed sensor to transmit the speed and mileage of the vehicle to the navigation computer.

- **speed/temp**
Using special sensors, Lowrance units can digitally display surface water temperature, boat speed

and distance log (odometer) in addition to showing the depth range. Serious anglers, especially tournament pros, use speed and temperature information as aids in their fishing strategies. Note: displaying speed is used solely for controlling purposes, and does not work at "top-end" boat speeds.

- **spherical error probability (sep)**
The radius of a sphere, centred at the user's true location, that contains 50 percent of the individual three-dimensional position measurements made using a particular navigation system. A probability that a percentage of three-dimension measurements will lie within a sphere of given radius, with the sphere centered at truth or mean of the measurements; SEP specifies test cases for measurement errors of sensors of three dimensions, such as velocity east, north, and vertical.

- **spheroid**
Sometimes known as ellipsoid; a perfect mathematical figure which very closely approximates the geoid. Used as a surface of reference for geodetic surveys; A body that is shaped like a sphere but is not perfectly round.

- **spillover**
Satellite signal that falls on locations outside the beam pattern's defined edge of coverage.

- **spin modulation**
Periodic amplitude variations resulting from the rotation of a satellite's antennas about its axis, rotating the antenna peaks-and-nulls.

- **spin stabilization**
A form of satellite stabilization and attitude control which is achieved through spinning the exterior of the spacecraft about its axis at a fixed rate.

- **spinaxis**
The axis of rotation of a gyroscope.

- **s-plane**
Continuous complex frequency plane; S-plane is used in control systems engineering in the design of control laws.

- **splitter**
A passive device (one with no active electronic components) which distributes a television signal carried on a cable in two or more paths and sends it to a number of receivers simultaneously.

- **spoiler**
A control surface on fixed-wing aircraft, usually mounted to the wings, that provides roll control and lift; Symbols: delta sub S; Typical Units: rad, deg.

- **spoofing**
The deliberate transmission of fake signals to skew the position

calculations of a GPS receiver. The spoofer mimics a GPS satellite, rather like a pseudolite, but with disruptive intent; The process of replicating the GPS code so that the user computes incorrect position solutions; The deliberate transmission of fake signals to skew the position calculations of a GPS receiver. The spoofer mimics a GPS satellite, rather like a pseudolite, but with disruptive intent.

- **spot**
  Systeme Pour l'Observation de la Terre. French, polar-orbiting Earth observation satellite(s) with ground resolution of 10 meters. SPOT images are available commercially and are intended for such purposes as environmental research and monitoring, ecology management, and for use by the media, environmentalists, legislators, etc.

- **spot beam**
  A focused antenna pattern sent to a limited geographical area. Spot beams are used by domestic satellites to deliver certain transponder signals to geographically well defined areas such as Hawaii, Alaska and Puerto Rico; The focusing of energy from a satellite to the ground to concentrate the limited available effective radiated power (EIRP).

- **spot image**
  Company that markets data gathered by the SPOT satellite worldwide.

- **spread spectrum**
  The received GPS signal is wide bandwidth and low power. The L-band signal is modulated with a pseudo-random noise code to spread the signal energy over a much wider bandwidth than the signal information bandwidth. This provides the ability to receive all satellites unambiguously and to give some resistance to noise and multipath; A system in which the transmitted signal is spread over a frequency band much wider than the minimum bandwidth needed to transmit the information being sent. This is done by modulating with a pseudo random code, for GPS. A signal normally requiring a narrow transmission bandwidth but spread over a much larger bandwidth. For example, the 50-bits-per-second GPS navigation message, which could be transmitted in a bandwidth of 50 Hz or so, is spread over a bandwidth of about 1 MHz by the C/A-code.

- **spread spectrum systems**
  A system in which the transmitted signal is spread over a frequency band much wider than the minimum bandwidth needed to

transmit the information being sent.

- **sps**
  Standard Positioning Service (civilian GPS).

- **squelch control**
  This control on a radio is used to silence the inherent background noise in a radio receiver so that only signals that appear above this background will be heard. To set a squelch control, turn it until the background noise is heard and then rotate the control until the noise just stops

- **sspa**
  Solid state power amplifier. A VSLI solid state device that is gradually replacing Traveling Wave Tubes in satellite communications systems because they are lighter weight and are more reliable.

- **stabilization**
  Maintenance of a desired orientation of a vehicle or device with respect to one or more reference directions.

- **stabilizer**
  A control surface, usually mounted at end of the fuselage parallel to the wings, that provides pitch stability, some aircraft have an adjustable stabilizer; Symbols: delta sub H; Typical Units: rad, deg.

- **stable element**
  An instrument or device which maintains a desired orientation independently of the motion of the vehicle.

- **stable platform**
  A gimbal-mounted platform, usually containing gyros and accelerometers, whose purpose is to maintain a desired orientation in inertial space, independent of the motion of the vehicle.

- **standard positioning service (sps)**
  A positioning service made available by the United States Department of Defence which is available to all GPS civilian users on a continuous, worldwide basis (typically using C/A Code); The civilian Absolute Positioning accuracy obtained by using the pseudo-range data obtained with the aid of a standard single-frequency C/A-Code GPS receiver. Under "Selective Availability" the horizontal accuracy is stated to be 100m 2drms (or 95% of the time).

- **standard positioning system**
  The GPS capability which is available to all users, less accurate than the Precise Positioning System; SPS is the less accurate GPS capability which is freely available to anyone using a GPS receiver.

- **starfix**
  DGPS positioning system, specifically for use offshore. This

system is intended for the professional user and, in addition to a high degree of accuracy, is equipped with a wide range of data analysis and quality control possibilities.

- **state data**

  Data that defines aircraft parameters, such as position, velocity, attitude; Some standard terms for state data include: Groundspeed vector, wind speed vector, true airspeed vector, true bearing, true track, ground track angle, relative bearing, sideslip angle, drift angle, true heading, magnetic variation, gravitation earth speed vector, vertical velocity, air mass flight path angle, earth-referenced flight path angle, angle of attack, pitch, radar altitude, barometric altitude, earth radius, glideslope, gravity vector, lift vector, lateral acceleration vector, bank angle.

- **state-space model**

  A mathematical relationship of a system in time using state variables, inputs, outputs, and constants; The state-space model is composed of n state variables (x sub 1 , x sub 2 , ..., x sub n), m input variables (u sub 1 , u sub 2 , ..., u sub m), k output variables (y sub 1 , y sub 2 , ..., y sub k), and four constants a, b, c, and d. Alternatively, a state-space model can be expressed with matrices. Compare: continuous-time equation, difference equation, differential equation, discrete-time equation, Laplace transform, Z transform;

- **static gps**

  Carrier phase differencing technique where the integer ambiguities are resolved from an extended observation period through a change in satellite geometry.

- **static observations**

  A navigation satellite survey technique that requires roughly one hour of observation, with two or more receivers observing simultaneously, and results in high accuracy and vector measurements.

- **static positioning**

  Location determination when the receiver's antenna is presumed to be stationary on the earth. In the case of pseudo-range-based techniques this allows the use of various averaging techniques to improve the accuracy. Static Positioning is usually associated with GPS Surveying techniques, where the two GPS receivers are static for some observation period which may range from minutes to hours (and even in the case of GPS geodesy, several days); Positioning applications in which the positions of static or near static points are determined; The process of averaging GPS positions taken successively over a period of time with a stationary antenna to increase accuracy.

- **static pressure**
A measure of barometric pressure as if the sensor were not moving with respect to the air; Compare: total pressure; Symbols: p sub s; Typical Units: psi,lbf/in-squared; Dimensions: Mass /Time-squared * Length

- **station**
A structure on the ground, perhaps containing VOR or TACAN

- **stationkeeping**
Minor orbital adjustments that are conducted to maintain the satellite's orbital assignment within the allocated "box" within the geostationary arc.

- **statistic accuracy**
The industry standard definition of static accuracy is the standard deviation of error if a point set by a receiver was tested in a fix position for 24 hours.

- **status**
An indicator of how well a system or subsystem is working

- **status indicator**
An binary indicator of a particular aspect of a device; status indicators are independent of each other; status indicators listed in this dictionary are derived from existing programs.

- **status words**
Data words reported by devices to indicate status; Each bit is defined on a device-by-device basis. The number of words vary from device to device. Status words are used by maintenance personnel and maintenance software. Present, past, and test status words are reported.

- **statute mile**
A unit of length equal to 5,280 feet or 1,760 yards (1,609 meters) used in the U.S. and some other English-speaking countries.

- **steering screen**
Shows a graphic, "highway view" of the GPS user's Course Over Ground. Provides helpful instructions as to how far off-course...which direction to steer, right or left, to make corrections...and displays related nav-data pertaining to the waypoint.

- **stefan-boltzmann law**
States that radiant flux of a blackbody is equal to the temperature to the fourth power times the Stefan-Boltzmann constant.

- **steppe**
A vast semi-arrid grassy plain. The steppe consists of three vegetation zones with significant differences in climate the wooded, or forest, steppe; the tillable steppe, or prairie; and the non-tillable steppe.

- **stereo model**
  Three-dimensional visual impression produced by viewing a pair of overlapping images through a stereoscope.

- **stereo pair**
  Two overlapping images or photographs that may be viewed stereoscopically.

- **stereopsis**
  The ability for objects to be perceived in three dimensions as a result of the parallax differences produced by the eye base.

- **stereoscope**
  Binocular optical device for viewing overlapping images or diagrams. The left eye sees only the left image, and the right eye sees only the right image.

- **stop-and-go positioning**
  This is a GPS Surveying "high productivity" technique which is used to determine centimetre accuracy baselines to static points, using site observation times of the order of 1 minute. Only carrier phase that has been converted into unambiguous "carrier pseudo-range" is used, necessitating that the ambiguities be resolved BEFORE the survey starts (and again at any time the satellite tracking is cut, e.g. due to signal obstructions). It is known as the "stop & go" technique because the coordinates of the receiver are only of interest when it is stationary (the "stop" part), but the receiver continues to function while it is being moved (the "go" part) from one stationary setup to the next. As the receiver must track the satellite signals at all times, hence the transport of the receiver from one static point to another must be done carefully.

- **stop-and-go surveying**
  A precision DGPS surveying technique, also referred to as semi kinematic, by which the rover is used to collect data a point for a few seconds or minutes before moving on.

- **straight line navigation**
  The standard method of navigation used by recreational GPS products. When commanded to navigate to a waypoint, the unit draws a straight, dotted line from the present position to the selected waypoint. It's the shortest, most direct route to the destination. Caution: Straight line navigation does not take into account any obstacles in the path; interim waypoints may be required to navigate safely around obstacles.

- **strap-down inertial navigation equipment**
  Inertial navigation equipment wherein the inertial sensors (for example, gyros and accelerometers) are directly mounted to the vehicle (eliminating the stable platform

and gimbal system) to sense the linear and angular motion of the vehicle. In this equipment, a computer uses gyro information to resolve the accelerations that are sensed along the carrier axes, and to refer these accelerations to an inertial frame of reference. Navigation is then accomplished in the same manner as in systems using a stable platform.

- **stratosphere**

Region of the atmosphere between the troposphere and mesosphere, having a lower boundary of approximately 8 km at the poles to 15 km at the equator and an upper boundary of approximately 50 km. Depending upon latitude and season, the temperature in the lower stratosphere can increase, be isothermal, or even decrease with altitude, but the temperature in the upper stratosphere generally increases with height due to absorption of solar radiation by ozone.

- **structure**

Structure is basically all solid objects rising from the bottom of a lake or river that isn't part of the actual bottom...like sunken trees, brush and rock piles...which creates habitats, feeding grounds and cover for game fish. For locating and separating close-holding fish from structure, GRAYLINE® is the preferred choice of more anglers.

- **sts-n (synchronous transport signal -- level n)**

In SONET, a frame of N x 6480 bits that is transported at N x 51.84 Mb/s. STS-N signals are created by interleaving N STS-1 signals, octet-by-octet. For various reasons, the values of N + 3, 12, 24, 48 and 96 are preferred. The optical equivalent OF STS Levels are expressed as OC-N levels.

- **subcarrier**

A second signal "piggybacked" onto a main signal to carry additional information. In satellite television transmission, the video picture is transmitted over the main carrier. The corresponding audio is sent via an FM subcarrier. Some satellite transponders carry as many as four special audio or data subcarriers whose signals may or may not be related to the main programming.

- **subsatellite point**

Point where a straight line drawn from a satellite to the center of the Earth intersects the Earth's surface.

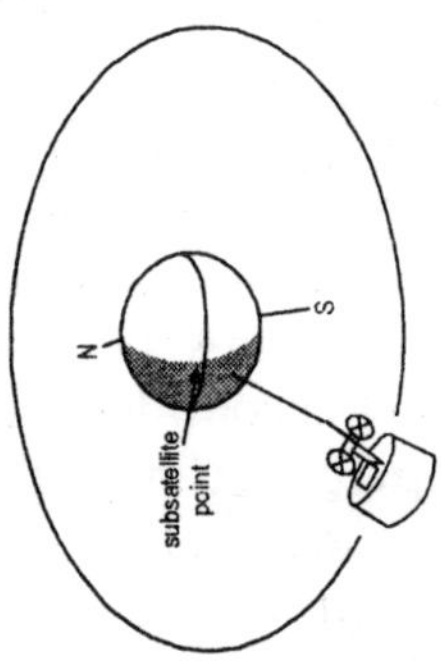

Point on the surface of the Earth directly between the satellite and the geocenter.

- **subscene**

  A portion of an image that is used for detailed analysis.

- **subscriber line charge (slc)**

  A monthly fee paid by telephone subscribers that is used to compensate the local telephone company for part of the cost of installation and maintenance of the telephone wire, poles and other facilities that link your home to the telephone network. These wires, poles and other facilities are referred to as the "local loop." The SLC is one component of access charges.

- **subsystem**

  1. A subunit of either the physical climate system (e.g., ocean dynamics) or the biogeochemical cycles (e.g., terrestrial ecosystems).
  2. A subunit of a spacecraft, e.g., the telemetry subsystem, the power subsystem, the sensor subsystem, etc.

- **subtractive primary colors**

  Yellow, magenta, and cyan. When used as filters for white light, these colors remove blue, green and red light, respectively.

- **sun**

  The closest star to Earth (149,599,000 km away on average). The sun dwarfs the other bodies in the solar system, representing approximately 99.86 per cent of all the mass in the solar system. One hundred and nine Earths would be required to fit across the Sun's disk, its interior could hold over 1.3 million Earths. The source of the Sun's energy is the nuclear reactions that occur in its core. There, at temperatures of 15 million degrees Celsius (27 million degrees Fahrenheit) hydrogen atom nuclei, called protons, are fused and become helium atom nuclei. The energy produced through fusion at the core moves outward, first in the form of electromagnetic radiation called photons. Next, energy moves upward in photon heated solar gas--this type of energy transport is called convection. Convective motions within the solar interior generate magnetic fields that emerge at the surface as sunspots and loops of hot gas called prominences. Most solar energy finally escapes from a thin layer of the Sun's atmosphere called the photosphere--the part of the Sun observable to the naked eye. The sun appears to have been active for 4.6 billion years and has enough fuel for another 5 billion years or so. At the end of its life, the Sun will start to fuse helium into heavier elements and begin to swell up, ultimately growing so large that it will swallow Earth. After a billion years as a "red giant," it will suddenly collapse into a "white dwarf." It may take a trillion years to cool off completely.

- **sunglint**
  Bright reflectance of sunlight caused by ripples on water.
- **sun-synchronous**
  Earth satellite orbit in which the orbit plane is nearly polar and the altitude is such that the satellite passes over all places on earth having the same latitude twice daily at the same local sun time; Describes the orbit of a satellite that provides consistent lighting of the Earth-scan view. The satellite passes the equator and each latitude at the same time each day. For example, a satellite's sun-synchronous orbit might cross the equator twelve times a day, each time at 3:00 p.m. local time. The orbital plane of a sun-synchronous orbit must also precess (rotate) approximately one degree each day, eastward, to keep pace with the Earth's revolution around the sun.
- **sun-synchronous orbit**
  a polar orbit where the satellite always crosses the Equator at the same local solar time.
- **supervised classification**
  Digital-information extraction technique in which the operator provides training site information that the computer uses to assign pixels to categories.
- **surface phenomenon**
  Interaction between electro-magnetic radiation and the surface of a material.
- **survey mode**
  Refers to observational emphasis upon frequent global coverage, usually with restricted spatial and spectral resolution, aimed at developing a consistent, long-term data product for later interpretation.
- **survey services**
  Services related to the measurement, management and mapping of locations, objects and operations, most of which involve a substantial navigation and positioning component.
- **sv**
  Satellite vehicle or space vehicle.
- **svg-scalable vector graphics**
  An XML markup language for describing two-dimensional vector graphics, both static and animated. It is an open standard created by the World Wide Web Consortium, which is also responsible for standards like HTML and XHTML.
- **swale**
  1. A low tract of land, especially when moist or marshy.
  2. A long, narrow usually shallow trough between ridges on a

beach, running parallel to the coastline.

3. A shallow trough like depression that carries water mainly during rainstorms or snow melts.

- **swashplate**

A device to control the pitch of rotors; on rotary-wing aircraft, the swashplate is controlled by a collective and a cyclic.

- **swath**

The area observed by a satellite as it orbits the Earth.

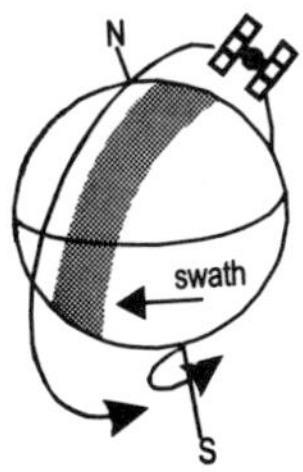

- **synchronization (sync)**

The process of orienting the transmitter and receiver circuits in the proper manner in order that they can be synchronized . Home television sets are synchronized by an incoming sync signal with the television cameras in the studios 60 times per second. The horizontal and vertical hold controls on the television set are used to set the receiver circuits to the approximate sync frequencies of incoming television picture and the sync pulses in the signal then fine tune the circuits to the exact frequency and phase.

- **synchronous transmission**

Digital transmission in which the time interval between any two similar significant instants in the overall bit stream is always an integral number of unit intervals.

- **synoptic chart**

Chart showing meteorological conditions over a region at a given time; weather map.

- **synoptic view**

The ability to see large areas at the same time.

- **synthetic aperture radar (sar)**

A high-resolution ground-mapping technique that effectively synthesizes a large receiving antenna by processing the phase of the reflected radar return. The along-track resolution is obtained by timing the radar return (time-gating) as for ordinary radar. The cross-track (azimuthal) resolution is obtained by processing the Doppler phase of the radar return. The cross-track "dimension" of the antenna is a function of the length of time over which the Doppler phase is collected. Radar system in which high azimuth resolution is achieved by storing and processing data on the Doppler shift of multiple return pulses in such a way as to give the effect of a much longer antenna.

- **synthetic stereo images**

Stereo images constructed through digital processing of a

single image. Topographic data are used to calculate parallax.

- **system**
  Applied to measurements, means the best value that the system can determine; Combination of components that constitute an imaging device.

- **systematic distortion**
  Geometric irregularities on images that are caused by known and predictable characteristics.

- **t1**
  The transmission bit rate of 1.544 millions bits per second. This is also equivalent to the ISDN Primary Rate Interface for the U.S. The European T1 or E1 transmission rate is 2.048 million bits per second. TDM digital channel carrier (1.544 MBPS).

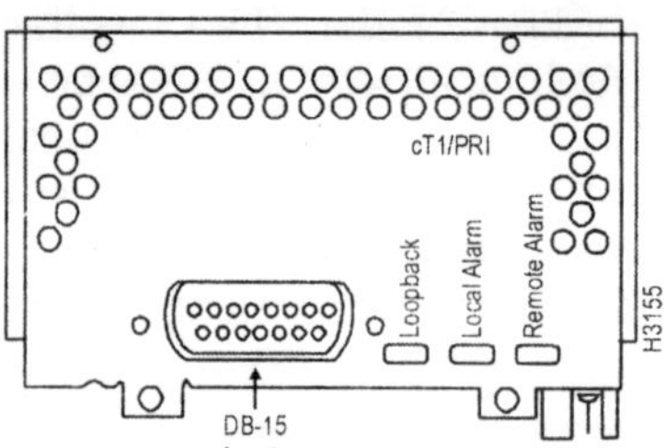

- **t3 channel (ds-3)**
  In North America, a digital channel which communicates at 45.304 Mbps.

- **tacan point-to-point (tcnp)**
  A basic guidance mode, providing lateral guidance to a point relative to a TACAN station by specified range and bearing;

- **tachometer**
  A device for measuring angular velocity.

- **tactical air navigation (tacan)**
  A basic guidance mode, providing lateral guidance to a TACAN station; Equipment that determines range and bearing to a radio station with a TACAN transmitter.

- **target**
  Object or point pointed by FLIR or radar; Object on the terrain of specific interest in a remote sensing investigation.

- **target cursor**
  A symbol on a display, moved by a track handle or similar device, to select objects on the display

- **tariff**
  The documents filed by a carrier describing their services and the payments to be charged for such services.

- **taut line**
  Basic guidance mode; provides vertical guidance to maintain a taut line for a dipping sonar, or other such device.

- **tcp/ip (transmission control protocol over internet protocol)**
  The de facto standard Ethernet protocols incorporated into

4.2BSD (Berkeley System Distribution-Version 4.2) Unix. TCP/IP was developed by DARPA (Defense Advanced Research Projects Agency) for internet-working and encompass both network layer and transport layer protocols. While TCP and IP specify two protocols at specific protocol layers, TCP/IP suite is often used to refer to the entire DoD (Department of Defense) protocol suite based upon these, including telnet (Remote Terminal Protocol), FTP (File Transfer Protocol) and UDP (User Datagram Protocol).

TCP (Transmission Control Protocol): The most common transport layer protocol used on the Ethernet and the Internet. TCP is built on top of IP (Internet Protocol) and is nearly always seen in the combination TCP/IP. It adds reliable communication, flow control, multiplexing and connection-oriented communication. It provides full-duplex, process-to-process connections.

IP (Internet Protocol): The network layer for the TCP/IP protocol suite widely used on Ethernet networks. IP is a connectionless, best-effort packet switching protocol. It provides packet routing, fragmentation and re-assembly through the data link layer.

- **tcp/ip suite**

Members of the Internet community share a family of protocols called TCP/IP. The name is shorthand for a suite of protocols whose major members are TCP (transmission control protocol), a transmission layer procedure, and IP (internet protocol), a network layer procedure. The five Layers of the TCP/IP suite of protocols are:

1. **Application Layer:** Interfaces user processes with lower level protocols.
2. **Transport Layer:** Contains protocols that establish, control and terminate network connections between data structure endpoints on source and destination hosts (TCP layer).
3. **Internet Layer:** Implements addressing that supports communication between network devices and provides routing for transmitting data between networks (IP layer).
4. **Datalink Layer:** Employs HDLC (High-level Data Link Control), LAP-B (Link Access Protocol-Balanced), LAP-D (Link Access Procedure on the D channel) and IEEE 802.2 protocols.
5. **Physical Layer:** Incorporates a wide range of physical interfaces such as EIA232/449 (Electronics Industry Association; Standard RS-232/449), ISDN (Integrated

Services Digital Network), IEEE 802 and ITU-T V.35 (International Telecommunications Union-Telecommunications Standards Sector; Standard V.35).

- **tdma**

Time division multiple access. Refers to a form of multiple access where a single carrier is the shared by many users. Signals from earth stations reaching the satellite consecutively are processed in time segments without overlapping.

- **tdop**

Time Dilution of Precision.

- **telecommunications relay service (trs)**

A free service that enables persons with TTYs, individuals who use sign language and people who have speech disabilities to use telephone services by having a third party transmit and translate the call.

- **telematics**

Combination of tele-communications and informatics.

- **telemeter**

To transmit data by radio or microwave links.

- **telemetry**

1. Telecommunications transmission to a distance of measured magnitude by radio or telephony with suitably coded modulation, e.g., amplitude, frequency, phase, pulse.
2. Transmission of data collected at a remote location over communications channels to a central station.
3. Surveying measurement of linear distances by use of tellurometer--a device that uses microwaves to measure distance.

Radio signals, originating at a satellite, that convey information on the performance or status of onboard subsystems. Also refers to the information itself.

- **telemetry, tracking and control (tt&c)**

Means of accessing the control systems in a spacecraft in order to position and configure the spacecraft; or the facility which monitors and controls the health of a satellite.

- **telephony**

Used to transmit sounds between widely removed points with or without connecting wires; The word used to describe the science of transmitting voice over a telecommunications network.

- **teleport**

Ground facility used to interface with a satellite during communications.

- **television and infrared observation satellite**

Television and Infrared Observation Satellite (TIROS): A series of NASA and NOAA satellites launched to monitor Earth's weather from outer space. The era of the meteorological satellites began with the launch of TIROS-1 on April 1, 1960. For the first time, it was possible to monitor weather conditions over most of the world regularly from space.

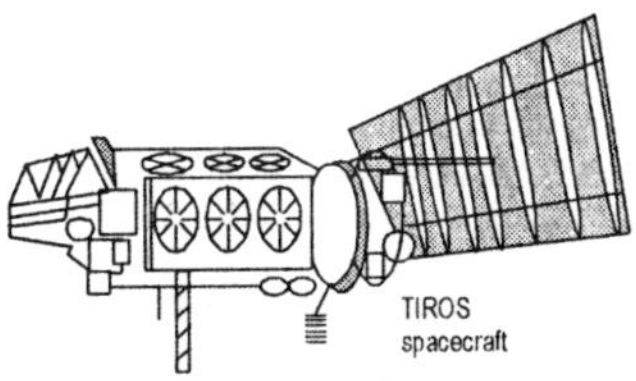

A series of these satellites were launched throughout the 1960s, those funded by NASA for research and development were called TIROS, and those funded by the Environmental Science Services Administration (ESSA, the predecessor of NOAA) for the operational system were called ESSA. A second generation of ITOS/NOAA environmental satellites was initiated by the launch of ITOS-1 in 1970, followed by a number of NOAA satellites. The third generation of TIROS-N/NOAA environmental satellites was initiated by the launch of TIROS-N in 1978. Pairs of acronyms such as ITOS/NOAA arise because NASA funds and names its prototype satellites and then the operating agency funds and names the rest of the series.

- **television receive only (tvro)**

A satellite receiving antenna, also known as a downlink or a backyard dish.

- **temperature**

A measure of the energy in a substance. The more heat energy in the substance, the higher the temperature. The Earth receives only one two-billionth of the energy the sun produces. Much of the energy that hits the Earth is reflected back into space. Most of the energy that isn't reflected is absorbed by the Earth's surface. As the surface warms, it also warms the air above it.

- **temporal decorrelation**

The age or time lapse in corrections used in differential GPS. The longer the time lapse between the corrections, the less accurate they become

- **terabit**

A trillion (1012) bits.

- **terminal node controller (tnc)**

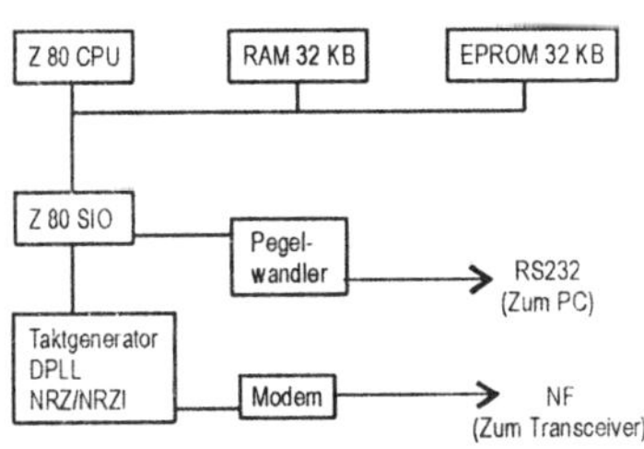

A device that converts and controls the radio transmission and reception of digital signals from a computer.

- **terrain avoidance (ta)**

Flight such that the aircraft maintains a constant barometric altitude but flies around obstacles.

- **terrain following (tf)**

A basic guidance mode, providing vertical guidance to maintain an operator selected radar altitude above the terrain. Flight such that the aircraft tries to maintain a constant height above the terrain, usually in the range of 100-1,000 ft; Uses a g-command from the Multi-Mode Radar to generate a flight director cue. This controls the aircraft flight path so that the set clearance altitude is achieved over major high points in the terrain with zero flight path angle.

- **terrain referenced navigation (trn)**

A navigation mode based on comparison of barometric altitude and radar altitude with a map; Kalman filters correlate the terrain data and the altitudes. A primary navigation data source. TRN combines INS with map references. It is most accurate over rough terrain. Does not give accurate data while over flat areas or water.

- **terrestrial**

In communications parlance, this refers to Earth, land or ground-based telecommunications systems.

- **test status words**

Status words reported at the conclusion of a test. Values: test in progress, not tested, normal, redundant, degraded, failed;

- **texture**

Frequency of change and arrangement of tones on an image. the gps (global positioning system) is a "constellation" of 24 well-spaced satellites that orbit the earth and make it possible for people with ground receivers to pinpoint their geographic location. the location accuracy is anywhere from 100 to 10 meters for most equipment. accuracy can be pinpointed to within one (1) meter with special military-approved equipment. gps equipment is widely used in science and has now become sufficiently low-cost so that almost anyone can own a gps receiver. the gps is owned and operated by the u.s. department of defense but is available for general use around the world. briefly, here's how it works:

1. 21 gps satellites and three spare satellites are in orbit at 10,600 miles above the earth. the satellites are spaced so that from any point on earth, four satellites will be above the horizon.
2. each satellite contains a computer, an atomic clock,

and a radio. with an understanding of its own orbit and the clock, it continually broadcasts its changing position and time. (once a day, each satellite checks its own sense of time and position with a ground station and makes any minor correction.)

3. on the ground, any gps receiver contains a computer that "triangulates" its own position by getting bearings from three of the four satellites. the result is provided in the form of a geographic position - longitude and latitude - to, for most receivers, within 100 meters.
4. if the receiver is also equipped with a display screen that shows a map, the position can be shown on the map.
5. if a fourth satellite can be received, the receiver/ computer can figure out the altitude as well as the geographic position.
6. if you are moving, your receiver may also be able to calculate your speed and direction of travel and give you estimated times of arrival to specified destinations. the gps is being used in science to provide data that has never been available before in the quantity and degree of accuracy that the gps makes possible. scientists are using the gps to measure the movement of the arctic ice sheets, the earth's tectonic plates, and volcanic activity. gps receivers are becoming consumer products. in addition to their outdoor use (hiking, cross-country skiing, ballooning, flying, and sailing), receivers can be used in cars to relate the driver's location with traffic and weather information.

- **thematic mapper (tm)**

A cross-track scanner deployed on Landsat that records seven bands of data from the visible through the thermal IR regions; A Landsat multispectral scanner designed to acquire data to categorize the Earth's surface. Particular emphasis was placed on agricultural applications and identification of land use. The scanner continuously scans the surface of the Earth, simultaneously acquiring data in seven spectral channels. Overlaying two or more bands produces a false color image. The ground resolution of the six visible and shortwave bands of the Thematic Mapper is 30 meters, and the resolution of the thermal infrared band is 120 meters. Thematic mappers have been flown on Landsats-4 and -5.

- **thermal conductivity (k)**

Measure of the rate at which heat will pass through a material, expressed in calories per centimeter per second per degree Centigrade.

- **thermal crossover**
  On a plot of radiant temperature versus time, the point at which temperature curves for two different materials intersect.
- **thermal diffusivity (k)**
  Governs the rate at which temperature changes within a substance, expressed in centimeters squared per second.
- **thermal inertia (p)**
  Measure of the response of a material to temperature changes, expressed in calories per square centimeter per square root of second.
- **thermal infrared**
  Electromagnetic radiation with wavelengths between about 3 and 25 micrometers.
- **thermal ir**
  IR region from 3 to 14 m that is employed in remote sensing. This spectral region spans the radiant power peak of the earth.
- **thermal ir image**
  Image acquired by a scanner that records radiation within the thermal IR band.
- **thermal ir multispectral scanner (tims)**
  Airborne scanner that acquires multispectral images within the 8-to-14mm band of the thermal IR region.
- **thermal model**
  Mathematical expression that relates thermal and other physical properties of a material to its temperature. Models may be used to predict temperature for given properties and conditions.
- **thermocline**
  These are areas underwater where warmer layers of water meet cooler layers, and where fish are often active. Generally, baitfish hang just above the thermocline, while larger game fish are found suspended in or just below it. Lowrance units will display thermoclines when used in the manual mode, showing actual sonar returns, and with the sensitivity set correctly.
- **thermography**
  Medical applications of thermal IR images. Images of the body, called thermograms, have been used to detect tumors and monitor blood circulation.
- **threat avoidance (tha)**
  Flight cues designed to avoid enemy threats, such as anti-aircraft artillery and aircraft.
- **three-dimensional (3d) navigation**
  Navigation mode in which altitude and horizontal position are determined from satellite range measurements.

- **three-dimensional coverage**
The number of hours-per-day when four or more satellites are available with acceptable positioning geometry. Four visible satellites are required to determine location and altitude.

- **throttle**
A flight control operated by moving fore or aft with hands, primarily to control thrust (speed) in fixed-wing aircraft.

- **throttle cue**
A longitudinal flight director cue for fixed-wing aircraft, primarily to control speed, by changing power.

- **thrust**
Force, created by engines and rotors, acting in the direction of the engine; Symbols: T; Typical Units: lbf,kip; Dimensions: Mass * Length / Time-squared.

- **thunder**
The sound that results from lightning. Lightning bolts (static electricity) produce intense heat. This burst of heat makes the air around the bolt expand explosively, producing the sound we hear as thunder. Since light travels faster than sound, we see the lightning before we hear the thunder.

- **thunderstorm**
Local storm resulting from warm humid air rising in an unstable environment. Air may start moving upward because of unequal surface heating, the lifting of warm air along a frontal zone, or diverging upper-level winds (these diverging winds draw air up beneath them). The scattered thunderstorms that develop in the summer are called air-mass thunderstorms because they form in warm, maritime tropical air masses away from other weather fronts. More violent severe thunderstorms form in areas with a strong vertical wind shear that forces the updraft into the mature stage, the most intense stage of the thunderstorm. Severe thunderstorms can produce large hail, forceful winds, flash floods, and tornadoes.

- **ti - terrestrial interference**
Interference to satellite reception caused by ground based microwave transmitting stations.

- **tie-point**
A point on the ground which is common to two images. Several are used in the coregistration of images.

- **tilt**
(1) The angle which the antenna axis forms with the horizontal.
(2) The vertical angle between the axis of measurement and a reference axis; the reference is normally horizontal.

- **tilt rotor**
  An aircraft with tilting rotors for fixed-wing flight or rotary-wing flight.

- **time constant**
  Constant for a first-order filter determining time at which the output of the filter reaches nearly 0.6321 per cent of a step input; Symbols: tau; Typical Units: s; Dimensions: Time.

- **time division multiple access (tdma)**
  A method of accessing a satellite using time separation; sending simultaneous data streams interleaved using a clock to separate the various data packets.

- **time to first fix (ttff)**
  If you have not used your GPS unit for several months, the almanac data for the satellites may be out of date. The unit is capable of recollecting this information on its own, but the process can take several minutes. Time to First Fix (TTFF) is the time it takes a GPS receiver to find satellites after the user first turns it on (when the GPS receiver has lost memory or has been moved over 300 miles from its last location). The actual time required by a GPS receiver to achieve a position solution. This specification will vary with the operating state of the receiver, the length of time since the last position fix, the location of the last fix, and the specific receiver design.

- **time to go (ttg)**
  Digital reading showing the time remaining from your current position to the next waypoint. This function takes into account your Distance To Go (DTG), and your Velocity Made Good (VMG) to give you as closely as possible the amount of time left to reach your waypoint. Displayed in hours, minutes, and seconds, it will continue counting down until waypoint is reached.

- **tiros-n/noaa satellites**
  NOAA satellites that continuously orbit the Earth from North to South Pole (hence, polar orbiting) at an altitude of approximately 470 nautical miles (870.44 km or 540.86 statute miles). These environmental satellites collect visible and infrared imagery and provide atmospheric-sounding data and meteorological data relay and collection. A primary mission of TIROS-N/NOAA is to monitor the 70 per cent of the globe covered by water-where weather data is sparse and provide continuous data to the National Weather Service for use in numerical forecast modeling. Each TIROS-N/NOAA carries six primary systems:
  1. The Advanced Very High Resolution Scanning

Radiometer (AVHRR) senses clouds over both ocean and land, using the visible and infrared parts of the spectrum. It stores measurements on tape, and later plays them back to NOAA's command and data acquisition stations. The satellites also broadcast in real time, and the broadcasts can be received around the world by anyone equipped with a direct readout receiving station.

2. The TIROS Operational Vertical Sounder (TOVS) is a 3-part TIROS system to measure: a. Temperature profile of the Earth's atmosphere from the surface to 10 millibars; b. Water content of the Earth's atmosphere; c. Total ozone content of the Earth's atmosphere;
3. The ARGOS Data Collection and Platform Location System (DCS) collects data from sensors placed on fixed and moving platforms, including ships, buoys, and weather balloons, and transmits data to a ground station antenna. Because ARGOS also determines the precise location of these moving sensors, it can serve wildlife managers by monitoring and tracking the transmitters placed on birds and animals.
4. The Space Environment Monitor (SEM) measures energetic particles emitted by the sun over essentially the full range of energies and magnetic field variations in the Earth's near-space environment. Readings made by these instruments are invaluable in measuring the sun's radiation activity.
5. Search and Rescue Tracking (COSPAS/SARSAT) equipment receives emergency signals from persons in distress. The satellites transmit the signals to ground receiving stations. The signals then are forwarded to rescue coordination centers. The rescue centers compute the location of the signals and provide the coordinates of the emergency site (usually within a few miles).
6. Earth Radiation Budget Experiment (ERBE) is a radiometer, flown on NOAA 9 and 10, designed to measure all radiation striking and leaving the Earth. This enables scientists to measure the loss or gain of terrestrial energy to space. Shifts in this energy "budget" affect the Earth's average temperatures. Even slight changes can affect climatic patterns.

- **tmc (traffic message channel)**
  Digital channel used to provide silent, coded messages to in-vehicle applications in order to display route and traffic information in a user's native language.

- **token-ring local area network**
  A computer LAN arbitration scheme in which conflicts in the transmission of messages are avoided by the granting of "tokens" which give permission to send. A station keeps the token while transmitting a message, if it has a message to transmit, and then passes it on to the next station. Often, "Token Ring" is used to refer to the IEEE 802.5 token ring standard which is the most common type of token ring.

- **tone**
  Each distinguishable shade of gray from white to black on an image.

- **topex/poseidon**
  Ocean Topography Experiment, United States (NASA)/France (CNES). Launched in 1992, the mission carries a radar sensor--called an altimeter--to measure the ocean's surface topography with unprecedented precision. TOPEX/POSEIDON is a core element of the international World Ocean Circulation Experiment (WOCE) and the Tropical Ocean Global Atmosphere (TOGA) seagoing measurements program. Mission objectives are to:

  1. Study ocean circulation and its interaction with the atmosphere to understand climate change better;
  2. Improve our knowledge of heat transport in the ocean;
  3. Model global ocean tides;
  4. Study the marine gravity field;
  5. Calculate sea-level variations on both global and local scales.

- **topographic inversion**
  An optical illusion that may occur on images with extensive shades. Ridges appear to be valleys, and valleys appear to be ridges. The illusion is corrected by orienting the image so that the shadows trend from the top margin of the image to the bottom.

- **topographic map**
  A map that gives not only locations of roads, landmarks and other features, but also gives contour lines at different elevations to show valleys, hills and other topography of the area.

- **topographic reversal**
  A geomorphic phenomenon in which topographic lows coincide with structural highs and vice versa. Valleys are eroded on crests of anticlines to cause topographic lows, and synclines form ridge, or topographic highs.

- **topography**
  Relief of the land surface; the graphic portrayal of that relief in map form by the use of contour lines.

- **tornado**

A twisting, spinning funnel of low pressure air. The most unpredictable weather event, tornadoes are created during powerful thunderstorms. As a column of warm air rises, air rushes in at ground level and begins to spin. If the storm gathers energy, a twisting, spinning funnel develops. Because of the funnel's cloud and rain composition and the dust, soil, and debris it draws up, the funnel appears blackish in color. The most energetic storms result in the funnel touching the ground. In these tornadoes, the roaring winds in the funnel can reach 300 mph, the strongest winds on Earth. Funnels usually travel at 20 to 40 mph, moving toward the northeast. When tornadoes form over lakes or oceans they suck water into the funnel cloud and are called waterspouts.

- **torquing rate**

The angular rate at which the orientation of a gyro, w.r.t. inertial space, is changed in response to a command.

- **total ozone mapping spectrometer (toms)**

Flown on NASA's Nimbus-7 satellite, its primary goal is to continue the high-resolution global mapping of total ozone on a daily basis. The Nimbus-7 launch in 1978 enabled TOMS to begin delivering data in 1979 and continue providing information until 1993. TOMS has mapped the total amount of ozone between the ground and the top of the atmosphere, provided the first maps of the ozone hole, and continues to monitor this phenomenon. Because of its longevity, TOMS also has obtained information on the more subtle trends in ozone outside the ozone hole region. This results from development of a powerful new calibration technique that removes the instrument measurement drift that developed over the years. With this technique applied to the TOMS 14.5-year data record, a global ozone decrease of 2.69 per cent per decade was detected. To ensure that ozone data will be available through the next decade, NASA will continue the TOMS program using U.S. and foreign launches. In 1991, the former Soviet Union launched a Meteor-3 satellite carrying a TOMS instrument provided by NASA. A third TOMS will be launched onboard a NASA Earth probe satellite in 1994, and the Japanese Advanced Earth Observations Satellite (ADEOS) will carry a fourth TOMS when it launches in 1996.

- **total pressure**

A measure of barometric pressure in the moving air; Synonyms: dynamic pressure,

Pitot pressure, stagnation pressure; Compare: static pressure; Symbols: p sub t; Typical Units: psi,lbf/in-squared; Dimensions: Mass /Time-squared * Length

- **tow**

Time on week in seconds from midnight Saturday in UTC.

- **township**

A unit of survey of the public lands of the United States, normally a quadrangle approximately 6 miles on a side with boundaries conforming to meridians and parallels within established limits, containing 36 sections.

- **tracback**

The proprietary Garmin feature which takes your current track log and converts it into a route to guide you back to a starting position.

- **track**

(1) The resultant direction of actual travel projected in the horizontal plane and expressed as a bearing.

(2) The component of motion that is in the horizontal plane and represents the history of accomplished travel.

A planned or intended horizontal path of travel with respect to the Earth rather than the air or water. The track is expressed in degrees from 000° clockwise through 360° (true, magnetic, or grid).

- **track angle error (tae)**

A guidance control law parameter, generated by the lateral guidance modes; Typical Units: rad.

The angle, at any moment, between the track of an aeroplane and the Meridian, measured clockwise from 0 degrees to 360 degrees.

- **track handle**

A device to move a cursor in two axes on a display, much like a mouse.

- **track made good**

The single resultant direction from a point of departure to a point of arrival or subsequent position at any given time; may be considered synonymous with Course Made Good.

- **track up orientation**

Fixes the GPS receiver's map display so the current track heading is at the top of the screen. trackingThe process of following a moving object or a variable input quantity. This process may be carried out manually or automatically. In radar, target tracking angle, range, or GPS frequency is accomplished by keeping a beam or angle cursor on the target angle, etc.

- **tracking and data relay satellite (tdrs)**

Geostationary satellite used to communicate between ground receiving stations and satellite such as Landsat.

- **tracking and data relay satellite system**

  Tracking and Data Relay Satellite System (TDRSS) An orbiting communications satellite, developed by NASA, used to relay data from satellite sensors to ground stations and to track the satellites in orbit.

- **trade winds**

  Surface air from the horse latitudes that moves back toward the equator and is deflected by the Coriolis Force, causing the winds to blow from the Northeast in the Northern Hemisphere and from the Southeast in the Southern Hemisphere. These steady winds are called trade winds because they provided trade ships with an ocean route to the New World.

- **trade-off**

  As a result of changing one factor in a remote sensing system, there are compensating changes elsewhere in the system; such a compensating change is known as a trade-off.

- **training area**

  A sample of the Earth's surface with known properties; the statistics of the imaged data within the area are used to determine decision boundaries in classification.

- **training site**

  Area of terrain with known properties or characteristics that is used in supervised classification.

- **transceiver**

  A combination transmitter and receiver.

- **transducer**

  A device, much like a microphone, that converts input energy of one form into output energy of another. Fish finders separate and enhance the information received from a transducer to show underwater objects; The part of the fish finder that functions like an antenna to send out sonar signals (sound waves) and receive return echoes that are displayed as pictures on the sonar screen. Lowrance transducers can be attached to the transom or inside the hull of a boat, or even to a trolling motor using a special mounting.

- **transfer alignment**

  A method of transfer of reference coordinates to an INS for initial alignment.

- **transient-free switch**

  A switch with a wash-out filter so that the output contains no transients (steps) at switch time.

- **transit**

  The satellite-based system that measured successive Doppler (frequency) shifts of signals transmitted from satellites in polar orbits to determine position; A precision surveying instrument; a theodolite in which the telescope can be reversed in direction by rotation about its horizontal axis.

- **translocation**
  A version of relative positioning which makes use of a known position, such as a USGS survey mark, to aid in the accurate positioning of a desired point. Here, the position of the mark, determined using GPS, is compared with the accepted value. The three-dimensional differences are then used in the calculations for the second point.

- **transmissivity**
  Property of a material that determines the amount of energy that can pass through the material.

- **transmitter**
  An electronic device consisting of oscillator, modulator and other circuits which produce a radio or television electromagnetic wave signal for radiation into the atmosphere by an antenna.

- **transparency**
  Image on a transparent photographic material, normally a positive image.

- **transpiration**
  Expulsion of water vapor and oxygen by vegetation.

- **transponder**
  A combination receiver, frequency converter, and transmitter package, physically part of a communications satellite. Transponders have a typical output of five to ten watts, operate over a frequency band with a 36 to 72 megahertz bandwidth in the L, C, Ku, and sometimes Ka Bands or in effect typically in the microwave spectrum, except for mobile satellite communications. Communications satellites typically have between 12 and 24 onboard transponders although the INTELSAT VI at the extreme end has 50.

- **transportation management satellite (tms)**
  The physical building and heart of the Dallas Area ITS system. Every piece of ITS field equipment is operated from the TMS facility. This facility also functions as the dispatch center for the Dallas District Courtesy Patrol.

- **transvision**
  Name given to the Fort Worth area Intelligent Transportation System (ITS).

- **travel time**
  In radar, the time interval between the generation of a pulse of microwave energy and its return from the terrain.

- **triangulation**
  A method of determining the location of an unknown point, as in GPS navigation, by using the laws of plane trigonometry.

- **trilateration**
  The same method as triangulation, except trilateration uses only distances for locating.

- **trim tab**

A secondary control surface, usually mounted to primary control surface such as aileron, elevator, rudder, or stabilizer, that controls the position of the primary control surface, and is controlled by the an operator or an autopilot; Symbols: delta sub T; Typical Units: rad, deg;

- **triple difference**

The arithmetic difference of sequential, doubly differenced carrier-phase observations. The triple-difference observable is free of integer ambiguities. It is a useful observable for determining initial, approximate coordinates of a site in relative GPS positioning and for detecting cycle slips in carrier-phase data

- **tristimulus colour theory**

A theory of colour relating all hues to the combined effects of three additive primary colours corresponding to the sensitivities of the three types of cone on the retina.

- **trivial baseline**

Trivial Baselines are those baselines formed when more than two GPS receivers are used simultaneously in the field to perform static GPS surveys. For example, when 3 receivers at points A, B, C are deployed only 2 baselines are independent (either A-B & A-C, AB & B-C, or AC & C-B), with the other one being trivial. This trivial baseline may be processed, but because the data used for this baseline has already been used to process the independent baselines, the baseline results should not be used for Network Adjustment or for quality control purposes unless the statistics (and variance-covariance matrix) are appropriately down weighted.

- **trop**

Tropospheric correction. The correction applied to the measurement to account for tropospheric delay. This value is obtained from the modified Hopfield model.

- **tropic of cancer**

An imaginary circle on the surface of the earth at 23.5°N latitude. The tropics are the two lines where the sun is directly overhead at noon on the two solstices - near June and December 21. The sun is directly overhead at noon on the Tropic of Cancer on June 21 (the beginning of summer in the Northern Hemisphere and the beginning of winter in the Southern Hemisphere) and the sun is directly overhead at noon on the Tropic of Capricorn on December 21 (the beginning of winter in the Northern Hemisphere and the beginning of summer in the Southern Hemisphere).

Cancer (cancer is Latin for crab), or Northern tropic is one of the five major circles of latitude that mark maps of the Earth. It is the parallel of latitude that lies currently 23° 26' 22" north of the Equator. It is the farthest northern latitude at which the sun can appear directly overhead, which occurs on the June solstice. North of this line is the subtropics and Northern Temperate Zone. The Tropic of Capricorn is at the opposite latitude south of the Equator. South of the Tropic of Cancer and north of the Tropic of Capricorn are the Tropics. It is difficult to specify an exact location for both tropic circles because the obliquity of the ecliptic varies between 21.5° and 24.5° with a 41,000 year periodicity.

- **tropical storm formation**

Tropical storms generally form in the eastern portion of tropical oceans and track westward. Hurricanes, typhoons, and willy-willies all start out as weak low pressure areas that form over warm tropical waters (e.g., surface water temperature of at least 80 degrees F). Initially, winds and cloud formations over the warm tropical waters are minimal. Both intensify with time. Formation of tropical storms also requires a significant Coriolis effect to induce proper spin in the wind formation. As the storm begins to organize itself into a coherent pattern, it will experience increased activity and intensity. When a storm develops a clearly recognizable pattern, it is referred to as a tropical depression. When wind speeds reach 35 knots (40.3 mph), it is called a tropical storm and is given a name. When wind speed equals or exceeds 74 mph, the storm is called a hurricane. In the western Pacific, a hurricane is referred to as a typhoon. In waters around Australia it is called a cyclone or willy-willy. Hurricanes intensify when moving over areas of increased water temperatures, and weaken over colder water surfaces. Upper atmosphere wind shear (different wind direction and speeds at different elevations) will frequently prevent or slow intensification of tropical storms by "spreading out" the storm horizontally and preventing the formation of strong updrafts of warm, humid air. Movement over a land-mass will weaken hurricane winds but will result in large-scale rain that can result in large-scale flooding. When encountering a strong frontal system (such as a polar front) the hurricane will curve and track along the leading edge of the front or become implanted in it. Satellite infrared imagery can identify surface water temperatures that will foster tropical storm development.

- **tropics**

The area between 23.5 degrees north and south of the equator.

This region has small daily and seasonal changes in temperature, but great seasonal changes in precipitation.

- **troposphere**

The lower atmosphere, to a height of 8-15 km above Earth, where temperature generally decreases with altitude, clouds form, precipitation occurs, and convection currents are active.

- **troposphere, tropospheric delay**

The Troposphere is the neutral atmosphere comprising the lower 8 km of the atmosphere. The Tropospheric Delay on GPS signals is of the non-dispersive variety because it is not frequency-dependent and hence impacts on both the L1 and L2 signals by the same amount (unlike that within the Ionosphere). The wet and dry components of the Troposphere cause the Delay to the signals, with the wet component be responsible for approximately 10% of the total delay. Various Tropospheric Delay models have been developed to estimate the Delay as a function of the satellite elevation angle, receiver height, and meteorological parameters such as temperature, pressure and humidity. The Delay in the Zenith direction is approximately 2.5m, increasing as the elevation angle of the satellite signal reduces. (This behaviour is described by the so-called Mapping Function, so that the Delay near the horizon is 3-5 times higher than in the Zenith direction.) The Delay is largely eliminated in Relative or Differential Positioning, however the residual Tropospheric Delay increases as the baseline length increases and may be a significant source of error (especially in the height component) for very high precision GPS Geodesy.

- **tropospheric emission spectrometer**

A high-resolution infrared spectrometer for monitoring the minor components of the lower atmosphere.

- **trough**

Elongated area of low atmospheric pressure, either at the surface or in the upper atmosphere.

- **true**

Best available estimate, such as true airspeed; referenced to true north, such as true heading;

- **true air speed**

The actual speed of an aircraft relative to the surrounding air; Airspeed corrected for instrumentation errors and air density; See Also: airspeed; Symbols: V sub 'TAS'; Typical Units: kt,ft/s; Dimensions: Length / Time.

- **true anomaly**

The angular distance, measured in the orbital plane from the earth's center (occupied focus) from the perigee to the current location of the satellite (orbital body).

- **true bearing**
  Bearing relative to true north; compass bearing corrected for compass error.

- **true colour image**
  An image in which the bands which most closely represent visible red, green, and blue have been assigned to red green and blue, thus producing an image which is similar in colour to what we see with the naked eye.

- **true course**
  Course relative to true north.

- **true heading**
  Heading relative to true north.

- **true north**
  The direction of the north pole from your current position. Magnetic compasses indicate north differently due to the variation between true north and magnetic north. A GPS receiver can display headings referenced to true north or magnetic north. It is different from Magnetic North, by the value of the Mag. Var.

- **ttff**
  Time To First Fix.

- **tty**
  A type of machine that allows people with hearing or speech disabilities to communicate over the phone using a keyboard and a viewing screen. It is sometimes called a TDD.

- **tune**
  To set the operating frequency or channel for a device.

- **turn (trn)**
  The degrees which must be added to or subtracted from the current heading to reach the course to the intended waypoint.

- **tvro-television receive only**
  Terminals that use antenna reflectors and associated electronic equipment to receive and process television and audio communications via satellite. Typically small home systems.

- **tweeking**
  The process of adjusting an electronic receiver circuit to optimize its performance.

- **two-dimensional (2d) navigation**
  Navigation mode in which a fixed value of altitude is used for one or more position calculations while horizontal (2D) position can vary freely based on satellite range measurements.

- **two-dimensional coverage**
  The number of hours-per-day with three or more satellites visible. Three visible satellites can be used to determine location if the GPS

receiver is designed to accept an external altitude input.

- **twt (traveling-wave tube)**

A microwave tube of special design using a broadband circuit in which a beam of electrons interacts continuously with a guided electromagnetic field to amplify microwave frequencies.

- **twta (traveling-wave-tube amplifier)**

A combination of a power supply, a modulator (for pulsed systems), and a traveling-wave tube, often packaged in a common enclosure.

- **uav (unmanned airborne vehicle)**

Unmanned autonomous mini-aircraft equipped with electro-magnetic measuring equipment Global Positioning System.

- **uhf/vhf automatic direction finding (u/v adf)**

An Automatic Direction Finder that determines relative bearing to a transmitter to which it is tuned, in either the UHF band or VHF band.

- **ultra high frequency (uhf)**

Radio frequencies in the band from 300 to 3,000 MHz; The band in the 500 to 900 MHz range, including TV channels 14 through 83.

- **ultraviolet radiation**

The energy range just beyond the violet end of the visible spectrum. Although ultraviolet radiation constitutes only about 5 per cent of the total energy emitted from the sun, it is the major energy source for the stratosphere and mesosphere, playing a dominant role in both energy balance and chemical composition. Most ultraviolet radiation is blocked by Earth's atmosphere, but some solar ultraviolet penetrates and aids in plant photosynthesis and helps produce vitamin D in humans. Too much ultraviolet radiation can burn the skin, cause skin cancer and cataracts, and damage vegetation.

- **umts (universal mobile telecommunication system)**

Standard for third-generation digital cellular telephony which enables users to receive video and audio on mobile phones; A transmission system for text, digitized voice, video, and multimedia for mobile computer and phone users no matter where they are located in the world.

- **unbundling**

The term used to describe the access provided by local exchange carriers so that other service providers can buy or lease portions of its network elements, such as interconnection loops, to serve subscribers.

- **undulation**
  The distance of the geoid above (positive) or below (negative) the mathematical reference ellipsoid (spheroid). Also known as geoidal separation, geoidal undulation, geoidal height.
- **unicast**
  A unicast application transmits a copy of every packet to every receiver.
- **unit functions**
  A collection of functions used as standard test cases in control systems engineering; The primary unit functions of interest in avionics are the unit impulse, the unit step, and the unit ramp. Symbols: u sub k (t).
- **unit impulse**
  A function used as a standard test case in control systems engineering; a spike of "area" one at time t = 0; Synonyms: impulse; See Also: unit functions; Symbols: u sub <0> (t), delta (t).
- **unit ramp**
  A function used as a standard test case in control systems engineering; a line of slope 1 starting at zero at time t = 0; Synonyms: ramp.
- **unit step**
  A function used as a standard test case in control systems engineering; a step from zero to one at time t = 0; Synonyms: step.
- **unitless**
  No units, such as ratios; a quantity with standard units of 1; a quantity with primary units of 1; Synonyms: dimensionless.
- **universal service**
  The financial mechanism which helps compensate telephone companies or other communications entities for providing access to telecommunications services at reasonable and affordable rates throughout the country, including rural, insular and high costs areas, and to public institutions. Companies, not consumers, are required by law to contribute to this fund. The law does not prohibit companies from passing this charge on to customers.
- **universal time**
  Local solar mean time at Greenwich Meridian. Some commonly used versions of Universal Time are:
  1. UT0: Universal Time as deduced directly from observations of stars and the fixed numerical relationship between Universal and Sidereal Time; 3 minutes 56.555 seconds.
  2. UT1: UT0 corrected for polar motion.
  3. UT2: UT1 corrected for seasonal variations in the earth's rotation rate.

4. UTC: Universal Time Coordinated; uniform atomic time system kept very close to UT2 by offsets. Maintained by the U.S. Naval Observatory. GPS time is directly relatable to UTC. UTC - GPS = seconds.

- **universal time coordinated (utc)**

Solar time in the Greenwich Mean Time (GMT), time zone but only ever adjusted in one-second increments. Because the GPS time is different to UTC a correction factor is applied to the received time that is reported by GPS receivers in UTC; UTC is the time as maintained by the U.S. Naval Observatory. Because of variations in the Earth's rotation, UTC is sometimes adjusted by an integer second. The accumulation of these adjustments compared to GPS time, which runs continuously, has resulted in an offset between GPS time and UTC. After accounting for leap seconds and using adjustments contained in the navigation message, GPS time can be related to UTC within 20 nanoseconds or better.

- **universal transverse mercator (utm)**

A system of fixed earth coordinates, sometimes used instead of latitude and longitude, accurate relative to others close by, such as ground troops.

- **unix**

An interactive time-sharing operating system (OS) invented in 1969 by Ken Thompson of Bell Labs. Twenty-two years after its invention, Unix had become the most widely used multi-user general-purpose operating system in the world. It was the first source-portable OS. Unix is now offered by many manufactures and is the subject of an international standardization effort with the Unix trademark being owned by X/Open.

- **unsupervised classification**

Digital information extraction technique in which the computer assigns pixels to categories with no instructions from the operator.

- **update**

A process by which position is reset with a known better position; updates occur at the data source level or at the physical device level, depending upon specific implementation of the device and the data source (choice of implementation is transparent above the data source); all updates are operator initiated, but the new position may be derived from other sources such as FLIR, MMR, or radio navigation; a basic control to a data source from controls and displays for updating a device.

- **update rate**

The GPS receiver specification which indicates the solution rate provided by the receiver when operating normally.

- **uplink**
  A transmission path by which radio or other signals are sent from the ground to an aircraft or a communications satellite.

- **upper atmosphere research satellite (uars)**
  UARS is part of a long-term, international program of space research into global atmospheric change. Beginning in 1991, NASA's UARS program began to carry out the first systematic, detailed satellite study of the Earth's stratosphere, mesosphere, and lower thermosphere; establish the comprehensive data base needed for an understanding of stratospheric ozone depletion; and bring together scientists and governments around the world to assess the role of human activities in atmospheric change. Launched on September 12, 1991, UARS became the first official space component of Mission to Planet Earth.

- **ups**
  Universal Polar Stereographic, a version of UTM.

- **ups-universal polar stereographic**
  An azimuthal projection that is used with Universal Transverse Mercator (UTM) projection systems to represent polar regions.

- **ura**
  User Range Accuracy: The contribution to the range-measurement error from an individual error source (apparent clock and ephemeris prediction accuracies), converted to range units, assuming that the error source is uncorrelated with all other error sources.

- **usat (ultra small aperture terminal)**
  Earth station satellite antenna with a diameter or cross-section dimension of about 50 cm or less.

- **user equivalent range error (uere)**
  Any error contributing to the error budget of stand-alone GPS receiver positioning, expressed as an equivalent error in the range between a user's receiver and a satellite. Also known as user range error (URE). UERE errors originate from different sources and thus are independent of each other. The total UERE is the square root of the sum of the squares of the individual errors. A prediction of maximum anticipated total UERE (minus ionospheric error) is provided in each satellite's navigation message as the user range accuracy (URA).

- **user interface**
  The way in which information is exchanged between the GPS receiver and the user. This takes

place through the screen display and buttons on the unit.

- **user range accuracy (ura)**
The contribution to the range-measurement error from an individual error source (apparent clock and ephemeris prediction accuracies), converted into range units, assuming that the error source is uncorrelated with all other error sources. Values less than 10 are preferred.

- **user segment**
The segment of the complete GPS system that includes the GPS receiver and operator.

- **usng-u.s. national grid**
A standard to create a more favorable environment for developing location-based services within the United States and to increase the interoperability of location services appliances with printed map products by establishing a nationally consistent grid reference system as the preferred grid for National Spatial Data Infrastructure (NSDI) applications. The U.S. National Grid is based on universally defined coordinate and grid systems and can, therefore, be easily extended for use world-wide as a universal grid reference system.

- **ut**
Universal time or GMT (Grenwich Mean TIme). It is different from UTC (difference can be almost 1 second). UT1 is the electronically distributed version.

- **ut0**
Universal Time as deduced directly from observations of stars (based on the rotation of .the earth) and the fixed numerical relationship between Universal and Sidereal Time; 3 minutes 56.555 seconds. When UT0 is corrected for the shift in longitude of the observing station caused by polar motion, the time scale UT1 is obtained.

- **ut1**
Is UT0 corrected for the effects of small movements of the earth relative to the axis of rotation (polar motion). Because Earth does not spin at exactly a constant rate, UT1 is not a uniform time scale.

- **ut2**
UT1 corrected for seasonal variation in the earth's rotation rate.

- **utc**
Universal Time Coordinated; uniform atomic time system kept very closely to UT2 by offsets. Maintained by the U.S. Naval Observatory. GPS time is directly relatable to UTC. UTC-GPS = 9 seconds (in 1994)

- **utm**
Universal Transverse Mercator. A type of map projection. The grid

lines you see on a USGS Quads represent a map projection known as Universal Transverse Mercator (UTM), and are 1000 meters apart; Each Zone in UTM is given a number from 00 to 59 starting at the International Date Line and progressing east. UTM also divides each zone horizontally as well. These divisions start at the equator and are 8 degrees wide. The first half of the alphabet is used for the Southern Hemisphere while the second half is used for the Northern Hemisphere. Thus a point just above the hemisphere would be the letter N proceeding to the letter X at the north pole, while just below it would be an M and proceeding backwards to the letter C at the south pole.

- **uv**
  Ultraviolet region of the electromagnetic spectrum ranging in wavelengths from 0.01 to 0.4m.

- **v.35**
  ITU-T standard describing a synchronous, physical layer protocol used for communications between a network access device and a packet network. V.35 is most commonly used in the United States and in Europe, and is recommended for speeds up to 48 Kbit/s.

- **value**
  descriptive information about a Feature. Values can be thought of as the answers to the questions posed by Attributes.

- **variable field**
  By NMEA standards, a data field which may or may not contain a decimal point and which may vary in precision following the decimal point depending on the requirements and the accuracy of the measuring device.

- **variance**
  A measure of the dispersion of the actual values of a variable about its mean. It is the mean of the squares of all the deviations from the mean value of a range of data.

- **variance factor**
  Reference variance, variance of unit weight A statistical measure of how close the observation residuals match the predicted errors. It is the square root of the sum of the weighted squares of the residuals divided by the degrees of freedom. If the errors in a network have been weighted correctly, the variance factor will approach 1.0.

- **vdop**
  Tertical (height only).

- **vector**
  Images defined by sets of straight lines that are defined by the locations of the end-points. Vector graphics require much less storage space than raster or bitmapped graphics.

- **vector format**
The expression of points, lines, and areas on a map by digitized Cartesian coordinates, directions, and values.

- **vector map**
Maps that use lines and polygons as opposed to pixels to display an image. Image resolution is sharp even when the image is displayed or printed at an enlarged perspective.

- **vegetation anomaly**
Deviation from the normal distribution or properties of vegetation. Vegetation anomalies may be caused by faults, trace elements in soil, or other factors.

- **vehicle detector**
Vehicle sensors either embedded in the pavement or mounted above the freeway to provide vehicle volume, speed, counts, headway, queue lengths, and vehicle classifications.

- **vehicle independent positioning (vip)**
Thales Navigation's VIP technology makes installation of the Magellan 750NAV, vehicle navigation GPS receiver, simple. The compact, transportable navigation computer and docking station offer a variety of mounting positions in most vehicles. Other systems require extensive wiring, which could critically reduce the vehicle's reliability. With VIP, the vehicle's position is computed by using feedback from 750NAV internal sensors to sense turns.

- **vehicle-derived navigation data**
Data obtained from measurements made at a vehicle.

- **velocity**
Rate of change of location, either scalar or vector, often with subscripts such as ENU or XYZ to denote the coordinate frame; time derivative of position; time integral of acceleration; Symbols: v, V; Typical Units: kt, ft/s; Dimensions: Length / Time.

- **velocity east**
Aircraft velocity in true east direction; Symbols: V sub E; Typical Units: kt, ft/s; Dimensions: Length /Time.

- **velocity error scale factor (kvsf)**
A guidance control law parameter, generated by the longitudinal guidance modes;

- **velocity made good (vmg)**
The rate of closure to a destination based upon your current speed and course.

- **velocity north**
Aircraft velocity in true north direction; Symbols: V sub N; Typical Units: kt, ft/s; Dimensions: Length / Time;

- **vertical**
Reference to ear th radial, for example, vertical velocity is velocity along earth radial.

- **vertical acceleration**
  Aircraft acceleration in earth vertical direction; Symbols: A sub V; Typical Units: ft/s-squared, g; Dimensions: Length / Time-squared.

- **vertical exaggeration**
  In a stereo model, the extent to which the vertical scale appears larger than the horizontal scale.

- **vertical guidance**
  Calculations for the vertical axis, rather than the longitudinal axis. The control law vertical axis input data are: Desired Altitude, Altitude Integral, Altitude Integral Gain, Altitude Integral Limit, Altitude Error Scale Factor, Magnitude limit for delta altitudes. The major output from Vertical Guidance is the altitude error for the selected vertical guidance mode.

- **vertical speed hold (vert spd)**
  A basic guidance mode, providing vertical guidance to maintain an operator selected vertical speed.

- **vertical velocity**
  Aircraft velocity in earth vertical direction; Symbols: V sub V; Typical Units: ft/s; Dimensions: Length / Time.

- **very high frequency (vhf)**
  The part of the radio spectrum from 30 to 300 megahertz, which includes TV Channels 2-13, the FM broadcast band and some marine, aviation and land mobile services.

- **very high-frequency omnidirectional range (vor)**
  A navigation aid operating at VHF and providing radial lines of position in any direction as determined by bearing selection within the receiving equipment.

- **vga (video graphics adapter)**
  A video display standard for computer monitors which evolved from the CGA (computer graphics adapter) and EGA (enhanced graphics adapter) standards of the early 1980's. In 1987, IBM introduced the PC-AT computer which adopted a new display standard called VGA which allowed computer monitors to show integrated lifelike pictures using multiple shades of color. This standard uses progressive line scanning.

- **vhf omnirange (vor)**
  Equipment that determines bearing to a radio station with a VOR transmitter; VOR transmitters usually contain ILS transmitters in addition to VOR; A basic guidance mode, providing lateral guidance to a VOR station. A VOR/ILS/MB. radio receiver provides a VHF Omni-Direction Range (VOR) function and Instrument Landing System with Marker Beacon (MB) function. VOR provides bearings to a fixed point (Localizer function); ILS/MB provides

course (localizer function) and glideslope deviations (Glide Slope (G/S) function) during approach to a runway, along with marker beacon indications. If VOR is collocated with a TACAN (VOR-TAC, or VORTAC), bearing and range information is provided.

- **video description**
An audio narration for television viewers who are blind or visually disabled, which consists of verbal descriptions of key visual elements in a television program, such as settings and actions not reflected in dialog. Narrations are inserted into the program's natural pauses, and are typically provided through the Secondary Audio Programming channel.

- **video transmission services (vts)**
The transmission of compressed or wideband video signals.

- **vidicon**
An imaging device based on a sheet of transparent material whose electrical conductivity increases with the intensity of EM radiation falling on it. The variation in conductivity across the plate is measured by a sweeping electron beam and converted into a video signal. Now largely replaced by cameras employing arrays of charge-coupled devices (CCDs).

- **vignetting**
A gradual change in overall tone of an image from the centre outwards, caused by the imaging device gathering less radiation from the periphery of its field of view than from the centre. Most usually associated with the radically increasing angel between a lens and the Earth's surface, and the corresponding decrease in the light-gathering capacity of the lens.

- **virtual circuit (vc)**
A communication arrangement in which data from a source user may be passed to a destination user over various real circuit configurations during a single period of communication.

- **virtual private network (vpn)**
A capability that allows users to create their own private information network.

- **visible radiation**
Energy at wavelengths from 0.4 to 0.7mm that is detectable by the human eye.

- **vissr- visible infrared spin**
Scan Radiometer carried by the GOES satellites.

- **visual dissonance**
The disturbing effect of seeing a familiar object in an unfamiliar setting or in an unexpected colour.

- **vmg-velocity made good**
Marine term Giving velocity to waypoint taking into effect the course you are steering.

- **vms (variable message signs)**
  Roadside message displays used for traffic control and information.

- **vog**
  Velocity Over Ground. Same as Ground Speed

- **vog-velocity over ground**
  The speed at which the GPS unit and the person operating it are moving with respect to the earth's surface (ignoring tidal and current activity). Same as Speed Over Ground (SOG) and ground speed.

- **volts**
  Electrical system voltage.

- **volume scattering**
  In radar, interaction between electromagnetic radiation and the interior of a material.

- **vortac**
  Collocation of VOR and TACAN providing distance and bearing to station; a basic guidance mode, providing lateral guidance to a set of a VOR station and a TACAN station that are collocated.

- **vsat**
  Very small aperture terminal. Refers to small earth stations, usually in the 1.2 to 2.4 meter range. Small aperture terminals under 0.5 meters are sometimes referred to Ultra Small Aperture Terminals (USAT's). Earth station satellite antenna with a diameter or cross-section dimension in the general range of 1.2 to 2.4 meters.

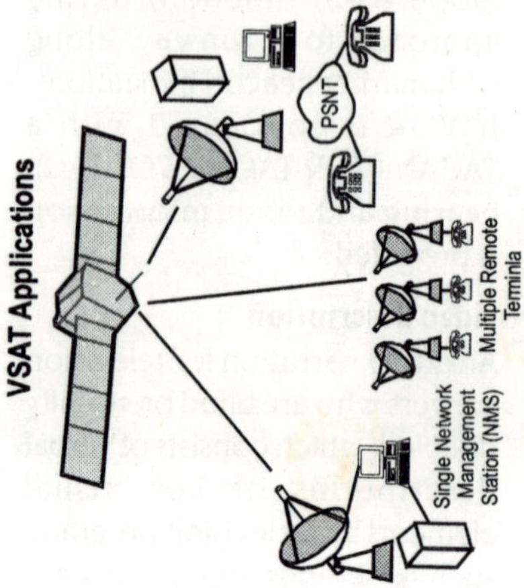

- **vss signal coupler**
  The VSS (vehicle speed sensor) wiring in Ford cars and trucks is different than that of other vehicles. This coupler amplifies the speed pulse in Fords, bringing it to a level that's compatible with a navigation unit.

- **vswr**
  Voltage Standing Wave Ratio. A measurement of mismatch in a cable, wave guide, or antenna system.

- **waas**
  Wide Area Augmentation System is a FAA funded system of equipment and software that augments GPS. The WAAS provides a satellite signal for WAAS users to support enroute and precision approach aircraft

navigation; WAAS is a network of 25 ground reference stations, ranging from the US to Canada to Mexico, that monitor GPS satellite information and create a GPS correction message. This correction message accounts for disturbances in the ionosphere, and for timing and satellite orbit errors. The WAAS-corrected signal can give your GPS unit a position accuracy of less than three meters (as compared to a typical 6-12 meters with the normal GPS signal). WAAS is ideal for open-water or flat terrain navigation, and is currently available in North America only.

- **wadgps**

A form of DGPS in which the user's GPS receiver receives corrections determined from a network of reference stations distributed over a wide geographical area. The corrections are typically supplied in real time by way of a geostationary communications satellite or through a network of ground-based transmitters

- **wage-wide area gps enhancement**

An attempt to improve GPS accuracy by providing more accurate satellite clock and ephemeris (orbital) data to specially-equipped receivers. WAGE is a method to increase the horizontal accuracy of the P"Y" Code receiver, by adding additional range correction data to the satellite broadcast navigation message. The navigation message for each satellite is updated once daily or as needed. With WAGE, this daily update of each satellite navigation message would contain the range corrections for every other satellite in the constellation. Thus, more timely range correction information would be available for each satellite, resulting in increased horizontal accuracy.

- **wap (wireless application protocol)**

Standard which brings Internet content to mobile phones.

- **warm start**

Used to describe state where a GPS receiver is powered on and still has valid almanac data available. This shortens the time required to locate satellites and arrive at the first position solution. The ability of a GPS receiver to begin navigating using almanac information stored in its memory from previous use

- **warning**

A signal which alerts the operator to a dangerous condition requiring immediate action (from MIL-STD-1472D); an annunciator that is the most critical (more than an advisory or a caution); Also, an indicator of potential failure soon;

Values: none, hot, low-power, high-power, other.

- **wash**
  1) Low or marshy ground washed by tidal waters.
  2) A stretch of shallow water.
  3) The dry bed of a stream (Western U.S.).

- **wash-out filter**

  A filter to smooth a transition due to change of input source, such as when changing modes.

- **waterproof**

  Most Garmin GPS units are waterproof in accordance with IEC 529 IPX7. IEC 529 is a European system of test specification standards for classifying the degrees of protection provided by the enclosures of electrical equipment. An IPX7 designation means the GPS case can withstand accidental immersion in one meter of water for up to 30 minutes. An IPX8 designation is for continuous underwater use.

- **watt (w)**

  Unit of electrical power equal to rate of work done by one ampere under a potential of one volt.

- **wave (figure)**
  1. In electricity, a periodic variation of an electric current or voltage.
  2. In physics, any of the series of advancing impulses set up by a vibration, pulsation, or disturbance in air or some other medium, as in the transmission of heat, light, sound, etc.

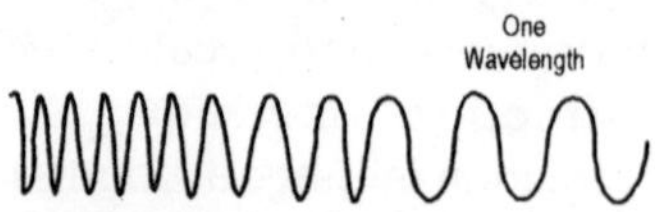

- **waveguide**

  A metallic microwave conductor, typically rectangular in shape, used to carry microwave signals into and out of microwave antennas.

- **wavelength**

  Distance between successive wave crests or other equivalent points in a harmonic wave.

- **waypoint**

  A (usually two-dimensional) coordinate that is input into a navigation device, such as a GPS receiver, representing a position that a vessel, aircraft, vehicle or person has to navigate to, with the aid of GPS (and/or any other position fixing device); Waypoints are locations or landmarks worth recording and storing in your GPS. These are locations you may later want to return to. They may be check points on a route or significant ground features. (e.g., camp, the truck, a fork in a trail, or a favorite fishing spot). Waypoints may be defined and stored in the unit manually by taking coordinates for the waypoint from a map or other reference. This can be done before ever leaving home. Or more usually, waypoints may be entered directly by taking a reading with the unit at the location itself, giving it a name, and then saving the point. Waypoints may also be put into the unit by referencing another waypoint already stored, giving the reference waypoint, and entering the distance and compass bearing to the new waypoint.

- **weather**

  Atmospheric condition at any given time or place. Compare with climate.

- **weather facsimile (wefax)**

  A system for transmitting visual reproductions of weather forecast maps, temperature summaries, cloud analyses, etc. via radio waves. WEFAX transmissions are relayed by NOAA's geostationary GOES spacecraft.

- **weather symbols**

  Some commonly used symbols are: weather terms:

  1. Clear - Sky cloud-free to 30 per cent covered.
  2. Sunny - Sunshine 70-100 per cent of the day. Partly sunny and partly
  3. Cloudy - Both terms refer to 30 to 70 per cent cloud cover. Partly sunny is used in the day; partly cloudy is used at night.
  4. Fog - A cloud on the ground Fog is composed of billions of tiny water droplets floating in the air.
  5. Snow - Precipitation of ice crystals.
  6. Snow flurries - Intermittent snowfall that may result in little accumulation.
  7. Sleet - Pellets of ice that form when rain or melting snowflakes freeze while falling. (Occurs in cold weather; hail usually occurs in summer.)
  8. Freezing rain - Rain that turns to ice on impact with the surface.
  9. Rain - Extended period of precipitation. Associated with large storm systems rather than single clouds or thunder storms.
  10. Showers - Brief interval of rain that does not affect a large area.
  11. Squall - Fast-moving thunderstorm or line of thunderstorms that often can produce damaging winds, hail, and tornadoes.

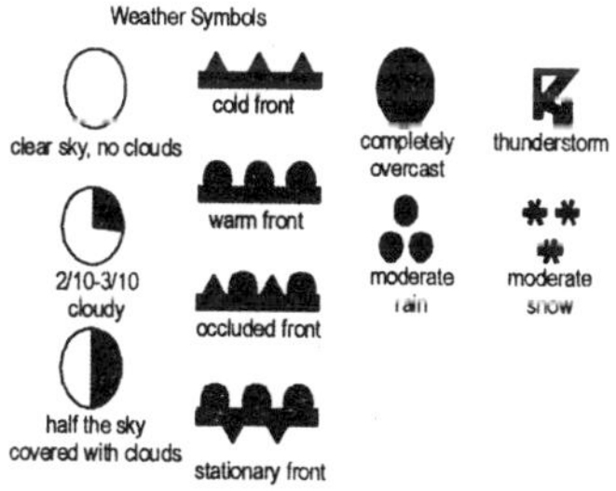

  12. Hail - Pieces of ice that fall from thunderstorms. Hail often

is composed of concentric rings of ice that form as the particle moves through "wet" and "dry" areas of the thunderstorm.

- **weather warning**

Statement that dangerous weather is likely or is occurring. Take action.

- **weather watch**

Statement about a particularly dangerous weather system that may occur at some specified time in the future.

- **weight on wheels (wow)**

Indication of whether the aircraft has weight on its wheels, meaning airborne or on the ground; weight on wheels can be detected by a sensor on the wheels, computed from other state data, or a combination.

- **wgs 84**

Geodetic datum for the GPS orbits and consequently positions and baselines computed using them. a = 6378137.000 m, b = 6356752.3100, 1/f = 298.257 223 563

- **wgs-72**

World Geodetic System (1972); the mathematical reference ellipsoid previously used by GPS, having a semimajor axis of 6378.135 km and a flattening of 1/298.26.

- **wheel**

A flight control operated by turning with hands in fixed-wing aircraft, primarily to control roll (heading) via the ailerons; wheel is connected to yoke.

- **wheel cue**

A lateral flight director cue for fixed-wing aircraft, primarily to control heading, by changing roll. where x is an observation of the estimated value μx.

- **wide area augmentation system (waas)**

A system of satellites and ground stations that provide GPS signal corrections for better position accuracy. A WAAS-capable receiver can give you a position accuracy of better than three meters, 95 per cent of the time. (At this time, the system is still in the development stage and is not fully operational.) WAAS consists of approximately 25 ground reference stations positioned across the United States that monitor GPS satellite data. Two master stations, located on either coast, collect data from the reference stations and create a GPS correction message.

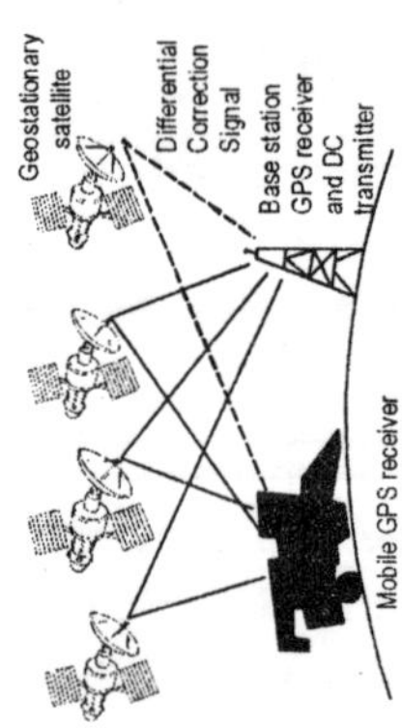

- **wide area dgps (wadgps)**
  Wide Area DGPS. A form of DGPS in which the user's GPS receiver receives corrections determined from a network of reference stations distributed over a wide geographical area. The corrections are typically supplied in real time by way of a geostationary communications satellite or through a network of ground-based transmitters

- **wide lane**
  A particular integer ambiguity value on one carrier phase range measurement or double difference carrier phase observation when the difference of the L1 and L2 measurements is used. It is a carrier phase observable formed by subtracting L2 from L1 carrier phase data: ?' = ?1 - ?2. The corresponding wavelength is 86.2 cm.

- **wideband**
  The property of any communications facility, equipment, channel or system in which the range of frequencies used for transmission is greater than 0.1% of the midband frequency. Wideband is often used to distinguish it from "narrowband", where both terms are subjectively defined relative to the implied context. The term also applies to the property of a circuit that has a bandwidth wider than normal for the type of circuit, frequency of operation or type of modulation.

- **wide-lane observable**
  The GPS observable obtained by differencing the carrier-phase observations simultaneously measured, in cycles, on the L1 and L2 frequencies. The effective wavelength is 86.2 centimeters. It can be useful in resolving carrier-phase ambiguities.

- **wien's displacement law**
  Describes the shift of the radiant power peak to shorter wavelengths as temperature increases.

- **wim (weigh in motion)**
  A technology for determining the weight of a commercial vehicle without requiring it to stop on a scale, using AVI to identify the vehicles and other technologies to measure the dynamic tire forces of a moving vehicle and then estimate the corresponding static weight.

- **wind**
  A natural motion of the air, especially a noticeable current of air moving in the atmosphere parallel to the Earth's surface. Winds are caused by unequal heating and cooling of the Earth and atmosphere due to absorbed, incoming solar radiation and infrared radiation lost to space--as modified by such effects as the Coriolis force, the condensation of

water vapor, the formation of clouds, the interaction of air masses and frontal systems, friction over land and water, etc. The preceding chart is an abbreviated version of the Beaufort Wind Scale, named for the British admiral who invented it in 1805.

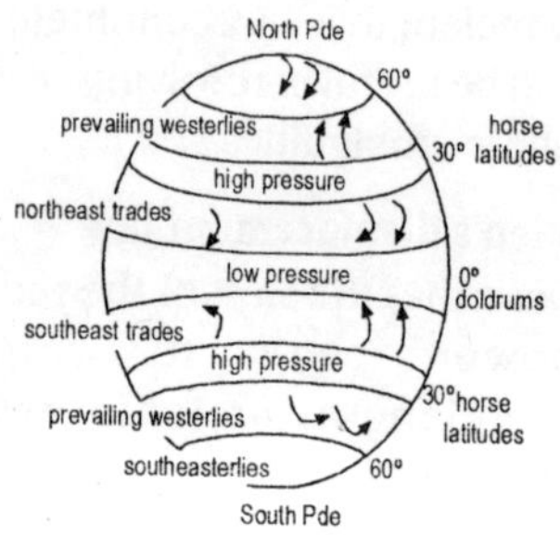

- **wind chill**

The wind can reduce significantly the amount of heat your body retains. The following wind chill chart does not take into account such variables as type of clothing worn, amount of exposed flesh, and physical condition, all of which would alter body heat.

- **wind vane**

An instrument used to indicate wind direction.

- **wind vector**

Arrow representing wind velocity. The arrow points in the direction of the wind. The length of the arrow is proportional to wind speed.

- **wind velocity**

Vector term that includes both wind speed and wind direction.

- **window**

Term used to denote a region of the electromagnetic spectrum where the atmosphere does not absorb radiation strongly. Overlap region between acquisition circles of two ground stations referenced to a specific satellite. Communications between two stations is possible when the sub-satellite point is within the window.

- **wireless cable**

A multichannel broadcast service where the multiple channels are transmitted with frequency division multiplex in a manner similar to cable television.

- **workstation**

A "smart" computer terminal that serves as a primary scientific research tool, offering direct access to experimental apparatus, information files, internal computers, and output devices, usually connected to an external communications network.

- **world geodetic survey 1984 (wgs 84)**

A common map datum reference. Geodetic datum for the GPS orbits and consequently positions and baselines computed using them

- **world ocean circulation experiment (woce)**

A study of the general global circulation of the oceans. It emphasizes the measurements and understanding needed to describe and understand the circulation, to simulate it, and to predict its changes in response to climatic changes.

- **world radio conference (wrc)**
  An international conference where standards and interference issues are discussed.The job of WRC is to review and revise the Radio Regulations and the international treaty governing the use of the radio-frequency.

- **wpt waypoint**
  A location saved in the receiver's memory which is obtained by entering data, editing data, calculating data or saving a current position. Used to create routes.

- **wwmx-world-wide media exchange**
  A centralized index of digital photos, where photos are tagged by the geographic location where they were shot.

- **x band**
  Radar wavelength region from 2.4 to 3.8 cm.

- **x.25**
  A set of packet switching standards published by the CCITT. x-bandA portion of the RF spectrum located between 7.25 GHz and 8.4 GHz which is dedicated to the United States Military for satellite communications.

- **xfcd**
  Extended Floating Car Data.

- **xte**
  Crosstrack Error. The distance you are off a desired course in either direction

- **xte cross track error**
  The perpendicular distance between the present position and the course line. Given as a distance, right or left, of course when facing the destination.

- **xte/xtk-crosstrack error**
  The lateral distance in either direction you are off the desired course.

- **xyz**
  A standard aircraft coordinate frame and sign convention, where nose, right wing, and down are positive X, Y, and Z, respectively; often used as subscripts.

- **y code**
  P Code that has been encrypted.

- **yaw**
  Angle of heading; Symbols: psi,Psi; Typical Units: rad, deg; Rotation of an aircraft about its vertical axis so that the longitudinal axis deviates left or right from the flight line.

- **yaw angle**
  (1) The horizontal angular displacement of the longitudinal axis of a vehicle from its neutral position

(2) The angle between a line in the direction of the relative wind and a plane through the longitudinal and vertical axes of the vehicle.

- **yaw rate**
  Rate of change of yaw; time derivative of yaw; Symbols: r; Symbols: psi dot; Typical Units: rad/s, deg/s; Dimensions: 1/Time.

- **y-code**
  An encrypted form of P-Code. Satellites transmit Y-Code in replace of P-Code when Anti-Spoofing is in effect; The term used to refer to the encrypted P-Code, generated within the satellites and transmitted on both the L1 and L2 carrier signals under the policy of "Anti-Spoofing". Civilian GPS receivers use proprietary signal processing techniques to make measurements of pseudo-range and carrier phase on both L-Band frequencies.

- **y-code tracking, civilian**
  Several methods of obtaining valid data from encrypted Y-code are available:
  1. Signal squaring (now obsolete) multiplies the signal by itself, thus deleting the carrier's code information and making distance measurement impossible. Carrier phase measurements can still be accomplished, although doubling the carrier frequency halves the wavelength, further weakening an already weak signal. This method requires collecting data over an extended period of time.
  2. Cross correlation is used where no local satellite navigation correction code is generated to match the L1 and L2 encrypted Y-codes. The ionosphere slows the L2 Y-code slightly in respect to the L1 Y-code creating a difference between the measurement of these distances. Once this difference is known, the frequencies can be matched and multiplied to leave pure carrier frequencies for measurement. This method removes the half-wavelength problem, but also results in a weakened signal that necessitates longer observation periods.
  3. Code correlation and squaring is a process wherein the L1 and L2 Y-codes are compared against a locally generated P-code. The difference in the signals is thus revealed, measured and squared so that pure carrier frequencies can be measured. Squaring once again weakens the resulting half-wavelengths of both carrier frequencies, and once again requires longer observation periods.

4. Thales Navigation's Z-Tracking is a unique method for retaining receiver lock and tracking capability. This technique separately matches the Y-code on L1 and L2 against a different, locally generated. P-code; a correlation that exposes the encrypting code on each frequency. Both carriers also contain the encrypting code, and so with sufficient signal integration, the encrypting signal is estimated for L1 and L2. Each calculation is fed to the other frequency, removing the encrypting code from each carrier frequency, which can then be purely measured.

- **yoke**

  A flight control operated by pushing and pulling with hands in fixed-wing aircraft, primarily to control pitch (altitude) via the elevators; yoke is mounted on a column between the operator's legs, positioned much like a steering wheel in a car; yoke control is achieved by pushing and pulling the wheel to move the column (yoke) fore and aft.

- **yoke cue**

  A vertical flight director cue for fixed-wing aircraft, primarily to control altitude, by changing pitch.

- **z count**

  Z-count is a 29-bit binary number consisting of the fundamental GPS time unit. The 10 most significant bits carry the GPS week number, while the 19 least significant bits give the Time Of Week (TOW) in units of 1.5 seconds.

- **z transform**

  A mathematical relationship to model a discrete function in the complex frequency domain (Z-plane); Z transforms are commonly used by systems engineers to describe avionics systems.

- **z-count**

  The GPS satellite clock time at the leading edge of the next data subframe of the transmitted GPS message (usually expressed as an integer number of 6 seconds); The fundamental GPS time unit. It is a 29-bit binary number, of which the 10 most significant bits give the binary representation of the GPS week number, and the 19 least significant bits give the time-of-week (TOW) count in units of 1.5 seconds.

- **zenith**

  The point on the celestial sphere that is directly above the observer.

- **zenith telescope**

  A telescope for observing stars near the zenith.

- **zero baseline**

  A Zero Baseline test can be used to study the precision of receiver measurements (and hence its correct operation), as well as the data processing software. The experimental setup, as the name implies, involves connecting two GPS receivers to the same antenna. When two receivers share the same antenna, biases such as those which are satellite (clock and ephemeris) and atmospheric path (troposphere and ionosphere) dependent, as well as errors such as multipath CANCEL during data processing. The quality of the resulting "zero baseline" is therefore a function of random observation error (or noise), and the propagation of any receiver biases that do not cancel in double-differencing.

- **zone of confusion (zoc)**

  A circular area centered at a TACAN station in which bearing is extremely noisy.

- **zoom**

  With zoom, you can enlarge the Lowrance sonar or GPS map display for a closer look at small detail and fish target signals. In the sonar automatic zoom mode, the bottom is constantly tracked and displayed in the lower portion of the screen (also called "zoom bottom track"). In the manual zoom mode, you can create a zoom "window" of any size, and move it anywhere up or down in the sonar water column...it doesn't remain locked on to display the bottom.

- **z-plane**

  Discrete complex frequency plane; Z-plane is used in control systems engineering in the design of control laws.

- **z-tracking**

  Thales Navigation's Z-Tracking process for mitigating or eliminating the effects of the U.S. Department of Defense Anti-Spoofing (AS) is a unique method for retaining receiver lock and tracking capability. This technique separately matches the Y-code on L1 and L2 against a different, locally generated P-code, a correlation that exposes the encrypting code on each frequency. Both carriers also contain the encrypting code, and so with sufficient signal integration, the encrypting signal bit is estimated for L1 and L2. Each calculation is fed to the other frequency, removing the encrypting code from each carrier frequency, which can then be purely measured.